Our Place, Time and Purpose in the Cosmos

Edward Tyler

This book is dedicated to those who throughout history have courageously pursued truth and the advancement of our awareness of the Cosmos, even when sometimes facing controversy or persecution.

Special thanks to Sheila Lehker for her love, support and editorial advice

CONTENTS

PREFACE

If you have ever flown in an airplane, with a window seat, and looked down on the Earth from about 30,000 feet above sea level, your experience may have challenged your perspective in regard to the circumstances or events in your life that seem critically important at the moment, or worries about things to come that consume your thoughts in the present. From this vantage point, it is clear that the world is a pretty big place, and those circumstances and worries that often feel overwhelming, are trivial in the greater scheme of the world. In the same sense, this book is about challenging our perspective, but on a much larger scale than the world as visible from an airplane.

While hidden during the day, the night sky reveals that there is far more to the Cosmos than we are aware of or that we typically consider in our daily lives. However, understanding what the night sky reveals can potentially transform our perspective with regard to what we see as important in our lives, and can impact how we interact with those around us. In the same way, what we learn can be transforming at the national level and can transform our outlook and interaction with other nations with whom we share this fragile spec of cosmic dust we call Earth.

This challenge of our perspective begins with an understanding of what is in the night sky, and where we live in relation to what we see. There is far more to the Cosmos than is apparent to our eyes when gazing up into the night sky. But we are also challenged through a better understanding of time in relation to those things we see in the night sky; realizing that our watches and calendars have little significance regarding time in the Universe, or in the greater Cosmos representing all that is or ever was. As with the mystery of what lies in and beyond our night sky, so too time in the cosmic perspective is itself not necessarily what we have come to expect and experience in our daily lives. Perhaps our perspective is most challenged in considering our purpose for existing in the context of the greater Cosmos. As we scratch beneath the surface of our everyday experience and consider our place and time in the Cosmos, we may discover a purpose for our lives. Moreover, we ask the question, are we alone in the Cosmos with this purpose.

Beyond this contemplation of our place, time and purpose in the Cosmos, this book considers our individual frames of reference by which we interpret what we observe in the night sky. Historically, such interpretations were largely driven by theologians pointing to Holy Scripture as the ultimate guide for understanding our world. During the scientific revolution beginning in the 16th century A.D., astronomers and mathematicians presented significant challenges to traditional interpretations of the Cosmos by the Church. The stage was set for contrasting perspectives and interpretations that have continued into contemporary society. Our minds are corralled from birth in many instances to adopt one frame of reference, or the other, to put our faith in theology, or to rely solely on our powers of observation and lean on our own understanding and interpretations of the Cosmos. There is a third possibility that is often overlooked when we recognize that no matter what philosophy we adopt, it does not alter what is actually true. While we may encounter a spectrum of philosophies in the world, there is only one truth. The third possibility is that both science and theology are one in the same, representing one truth for interpreting and understanding the Cosmos.

CHAPTER 1

HOME AMONG THE STARS

"Astronomy compels the soul to look upward, and leads
us from this world to another"

— Plato

"Eventually, we reach...the utmost limits of our telescopes.
There, we measure shadows, and we search among ghostly
errors of measurement for landmarks that are scarcely more
substantial."

— Edwin Hubble

If someone asks you where you live, you might reply with your
home address such as 4034 Place Avenue. Such a description of your
address would be accurate and, would presumably imply that you have
many neighbors whose own addresses would be identified with either
a lower or higher number.

You might also answer the question as to where you live by pointing
out the city, state or even the country in which you reside, such as
Richmond, Virginia, United States of America. And in this sense, you
would also presumably have many neighbors in other cities, and in
other countries around the Earth.

But look up at the night sky on a dark clear evening and ask yourself
again where do you, your neighbors, and everyone on Earth really live?
On the scale of the Cosmos, being inclusive of all there is or ever was,
and inclusive of whatever may lie beyond even our own Universe, what
is our cosmic address? As with a numeric address that implies others
reside nearby, so too our cosmic address is ultimately described by
identifying our cosmic neighbors.

EARTH FROM APOLLO 17 IN 1972

Our Nearest Neighbors

To figure out who our neighbors are on the cosmic scale, look up again at the night sky. Once you get past the wonder and inspiration that comes from looking up at all the stars in the heavens, you will notice a few stars in particular that are brighter than the others. If you were to watch these same stars night after night, you would find that they tend to wander across the sky relative to the other stars. The ancient Greeks called these *planētēs (from the Greek πλανήτης which means wanderer).* Today we simply call them the planets. By February 18, 1930 (discovery of Pluto), humans had discovered nine of these planets that revolve around our Sun, though Pluto was later demoted to the status of dwarf planet. These wanderers offer the first clue as to where our home lies in the Cosmos.

To understand where we are in relation to the planets, it is necessary to first talk about the units by which we measure the distance to these neighbors. Early measurement of distance was determined and communicated by using things in our everyday experience that we could compare to, such as measurements in feet being compared with the average size of a human foot. Historically in Europe, travel was measured in units of leagues, which represented the distance one could walk in an hour, or about 3 miles. Similarly, the unit of a mile originated in Rome from the term mille passus, which means a thousand paces, which measured 5,000 Roman feet. But in space, such distances, while accurate, are highly impracticable for effective communication. Our actual closest neighbor, the Moon, is approximately 240,000 miles, which equates to about 1,267,200,000 feet. Perhaps the simplest way to measure that distance is to use the Earth as our point of reference for distance. The Earth has a diameter of about 8,000 miles, so the distance to the Moon could easily be communicated as about 30 Earths away.

However, once you move away from the Moon, the distances involved to our nearest neighbors require another frame of reference for measurement. On the scale of our nearby cosmic neighborhood, the Solar System, a unit called an astronomical unit (AU) has been adopted to describe the distances between the planets. One AU represents the average distance from the Sun to the Earth or about 93,000,000 miles.

With that in mind, we again visit the wandering "stars", or planets, we observe in the night sky. In actuality, these stars are not so much wandering through the sky, so much as revolving around the Sun. Johannes Kepler (1571 A.D. to 1630 A.D.) was the first to work out mathematically precisely how the planets revolved around the Sun. In particular, Kepler's third law of planetary motion, based on the assumption of elliptical orbits, provides a mathematical correlation between the time a particular planet orbits the Sun, and the distance of that planet from the Sun. This mathematical relationship is presented below.

$$P^2 = a^3 \text{ (Kepler's third law of planetary motion)}$$

p = period of time in Earth years to revolve around the Sun
a = semimajor axis of the planet's orbit, or one half of the sum of the closest and furthest approach to the Sun in its orbit

Mercury. The first wanderer lies close to the Sun in the night sky, and takes only 88 days, or about 3 Earth months, to orbit, which based on Kepler's third law indicates that it lies about 0.39 AU (36,270,000 miles) from the Sun (technically this is actually the semimajor axis of its elliptical orbit, but it provides an approximation of the distance). This wanderer reveals itself in the early dawn and dusk hours near to the horizon and close to the setting or rising Sun. Because of its swift passage around the Sun, the ancient Romans named it after the messenger of the gods.

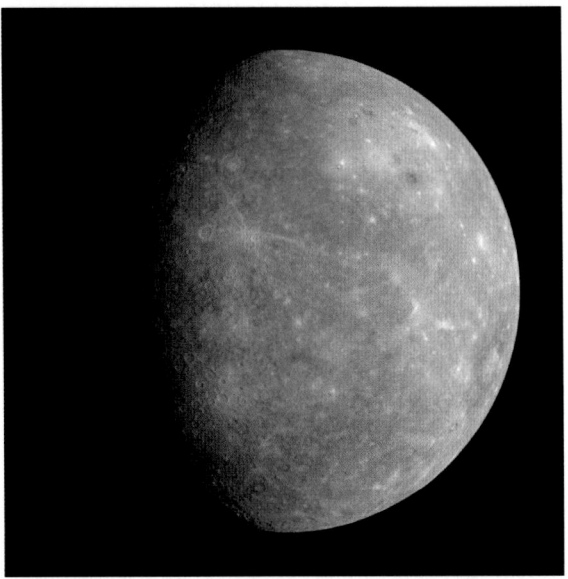

MERCURY

Imagine yourself in northern Arizona. This is a place characterized by rocky surfaces covered by layers of thin soil, and in the location where you find yourself, it is seemingly devoid of life. Imagine you are overlooking a large meteor crater left from an ancient impact event.

Suddenly, things begin to go wrong. You become aware that you feel as though you weigh almost nothing at all. Walking around with a new bounce in your step you feel as though you are about 40 percent of your former weight. Next, the atmosphere around you seems to get thinner and thinner, as though it was dissipating and being lost into space, leaving only trace gasses. Moreover, as it is a particularly hot day, the temperatures soar to an extreme of 800 degrees Fahrenheit, and this scorching hot day lasts for 88 days, or almost 3 months.

Eventually, the daylight retreats and gives way to darkness. Temperatures now plummet to minus 280 Fahrenheit. As with daylight, the night similarly last for about 88 days.

You are acutely aware, that you are no longer in northern Arizona, but on the planet Mercury. This is a planet only 3,032 miles in diameter, or about the width of the United States, that largely consists of a silicate surface overlying the largest iron core of any planet in the Solar System.

Moreover, Mercury is tidally locked with the Sun, and orbits the Sun twice, 176 days, and rotates on its axis three times, about 58 days each rotation, before completing a day and night cycle. With virtually no tilt in its axis relative to its orbit around the Sun, Mercury does not experience seasons. However, because of the low gravity and high heat experienced on the planet, almost no atmosphere is retained, except for trace amounts of hydrogen, helium, and oxygen. The absence of an appreciable atmosphere means that heat is not retained, and temperature extremes result between day and night.

Venus. This bright wanderer, named after the Roman goddess of love and beauty, rises in the evening or morning sky on Earth, and outshines all other objects in the night sky, except the Moon.

VENUS

This bright object seems constrained to be close to the Sun, though not as close as Mercury, and so is unable to travel across the night sky. With an orbital period of about 225 Earth days around the Sun, this wander lies about 0.72 AU from the Sun, the equivalent of about 66,960,000 miles.

Imagine you are in a diving bell, designed to withstand oceanic depths, and are lowered to a depth of 3,000 feet below sea level. At this depth, the "atmospheric" pressure is about 90 times that of the surface, such that every square inch of the diving bell is experiencing an external pressure of about 1,300 pounds.

Now imagine that the ocean around you slowly evaporates away while the pressure around you remains the same. Your instrumentation informs you that you are now surrounded by an oxygen-deprived atmosphere composed of about 96.5% carbon dioxide, 3.5% nitrogen, and lesser amounts of sulfur dioxide, argon, water vapor, carbon monoxide, helium, and neon. Further, your instrumentation tells you the temperature outside is 863 degrees Fahrenheit.

Looking out your small window, to the west you see the Sun beginning to rise, and as you gaze outwards you see that your diving bell is lying on solid reddish-brown ground, with an orange sky above. Looking around, you would observe a land surface, not like the bottom of the ocean, but rather a land characterized by volcanic mountains and highly weathered meteor craters. The land would be smoothed over as a result of a recent lava flow, or interaction with a corrosive atmosphere. The daylight would last seemingly forever and if you were keeping track, it's longevity would be the equivalent of about 117 Earth days. You would quickly realize that you were not in the depths of

SURFACE OF VENUS

Earth's ocean, or on Earth at all, but rather on the planet Venus. This planet is about the same size as the Earth, and is characterized by contrasts with the other planets, including a hotter temperature than any other planet. Moreover, Venus rotates in the opposite direction to the other planets, which explains the western sunrise.

Mars. Now consider another wanderer in the night sky. This one is not as bright the wanderer named after the goddess of love and beauty, but for the discerning eye this wanderer, with its reddish hue, has its own charm.

Not constrained to be close to the Sun, this wanderer is free to roam across the night sky, as if, as imagined by the ancient Romans who named it, the Roman god of war was being pulled across the sky in his chariot by his two dogs Phobos (fear) and Deimos (panic), the names of which were later ascribed to the two moons that orbit this red wanderer. With an orbital period of 1.88 years, this wanderer lies about 1.52 AU from the Sun or about 141,360,000 miles.

MARS

Imagine waking in the night, finding yourself entirely alone and outdoors in bitter cold temperatures. At first glance, a look up at the night sky might not seem remarkable or different from any previous night skies that you have looked upon, except for a particularly bright "star" near the horizon with a noticeably blue hue outcompeting the other stars for your attention. Now imagine looking to the horizon, as the sky begins to turn a beautiful pinkish red, forecasting the rising of the Sun. After a little while, the Sun would begin to rise, and the surrounding blue sky might not seem remarkable. However, if you were paying attention you would realize that a normal sunrise on Earth would be accompanied by a complement of red and orange. Later still into the day, the sky would turn a tawny color, a mixture of brown and orange. As you gaze around in this strange light, you would find yourself in a vast desert of reddish rusted soil that you would expect if iron predominated the soil, and there were sufficient oxygen and water vapor to cause its corrosion.

Now approaching mid-day, temperatures might reach up to 95 degrees Fahrenheit, and you might feel quite comfortable, if not a little warm. Suddenly you notice on the horizon a dust storm beginning to build, and then races in your direction. The storm seemingly covering the entire planet would hit with fierce winds upward of 100 miles per hour. Still, all these events, if you weren't paying close attention, might lead you to believe you woke up somewhere in the Sahara Desert. A day would come and go in a little more than 24 hours, and you would be none the wiser for the experience.

However, what happens next would lead you to a very different conclusion. Imagine feeling as if it is getting increasingly hard to breathe, as if the atmosphere itself is being stripped away steadily, which it has over time, because the planet you are on has no magnetic field to protect the atmosphere from getting stripped away by solar wind. After some time, you find yourself forced to don a space suit that you happen to have with you. The LCD display in your helmet tells you the atmosphere, what little there is, consists predominately of 96% carbon dioxide, with much smaller quantities of argon and nitrogen, trace amounts of oxygen, and water vapor. With this information, you decide to begin breathing through the supplied air that comes with the suit.

Now after all these events, you are thirsty and you begin walking north in a quest to satiate your thirst. As you begin your journey, you notice right away that you feel lighter, as though you have lost more than half your weight, which is attributable to the local gravity that for reasons unknown to you is about 38 percent of what you are accustomed to. You refocus in your quest for water and continue north. Unbeknownst to you, water lies far to the north in the ice caps of this lonely and desolate planet where you have found yourself. Your journey would take you about 3,000 miles, if you started at the equator, before encountering water. Even if you make it to the poles, the minus 225 Fahrenheit temperatures would be more than you can bear. By comparison, in his bid for the South Pole on Earth in the early 20th century, Earnest Shakelton and his companions, with the best cold weather gear available to them at the time, were able to tolerate temperatures as low as minus 100 degrees Fahrenheit, and only for a brief time.

In your quest, if you are lucky and not afraid of climbing a mountain or two, you might find water sooner. This planet turns out to have four seasons, though both a year and the seasons last twice as long as you are accustomed to on Earth. During the summer, occasional ice floes begins to move down steep mountainous slopes that might provide you with the precious resource you seek.

SURFACE OF MARS

At some point, the mystery would be unveiled, and you would realize you are not on Earth, but rather the planet Mars. This planet is only about one-half the Earths diameter, and at an average of 143 million miles from the Sun, it lies about 50 percent further from the Sun than Earth, which explains why the Sun appears smaller in the sky.

At this time, getting humans to Mars is mostly a question of whether we can, but not whether we must. The day will come when Earth's resources become sufficiently strained from the demand of a growing population that interplanetary habitation of Mars will become a necessity. At its closest approach to Earth in its orbit around the Sun, Mars is still about 50 million miles away. Traveling at speeds of perhaps 17,000 miles per hour, the trip will take at a minimum 123 days assuming your ship travels in a straight path and exactly intercepts Mars at its closest approach to Earth. When we get there, we will have become the Martians of so many science fiction novels.

Jupiter. Another wanderer, second only to Venus in its brightness in the night sky, was given its name from the Roman king of the gods. This is the wanderer first examined through a telescope by Galileo that turned humankind's understanding of the Universe upside down. With an orbital period of 11.86 years, this wanderer lies about 5.2 AU from the Sun or about 483,600,000 miles.

Imagine you are in a large hot air balloon floating through dark skies on a heavily overcast day. The skies, filled with water ice clouds, are as dark as you have seen, and below you is a vast dark ocean overlain by a dense fog seemingly extending forever into the horizon. Temperatures around you are a scorching 170 degrees Fahrenheit. As you glide perilously through the sky, lightning hundreds of times more powerful than you have ever experienced, strikes around you. A drizzle of rain, which turns out to be drops of helium and neon, falls from the dark skies above. An occasional clearing reveals dynamic displays of fantastic orange and brown cloud patterns higher in the atmosphere. During moments of particular clarity, daylight finds its way through the vast array of clouds, though the Sun's brightness is no more than about 4 percent of what you are accustomed.

JUPITER

You are floating in the atmosphere of Jupiter, a planet more than 10 times the diameter of the Earth and one-tenth the diameter of the

Sun, around which it orbits once every 12 Earth years. Despite having more than two and one-half times the mass of all the other planets combined, this massive gas giant rotates on its axis once every 10 hours, faster than any other planet in our Solar System.

The composition of Jupiter resembles that of a primordial star with 90 percent hydrogen and almost 10 percent helium. Moreover, because this gas giant is so large, internal pressures result in core temperatures of 43,000 degrees Fahrenheit, hotter than the surface of the Sun. Consequently, Jupiter gives off more heat than it receives from the Sun. It is theorized that if Jupiter were on the order of 80 times more massive, it would have auto-ignited and we would be living in a binary star system!

The extreme pressures and temperatures cause hydrogen to transition to a metallic state so that a vast "ocean" of metallic hydrogen surrounds the planet's core. In actuality, there is no clear transition from the metallic hydrogen "ocean" to the gaseous state of the lower atmosphere that consists of hydrogen, helium and trace amounts of methane, ammonia, and water. Water ice clouds are thought to predominate in the lower atmosphere that gives rise to thunderstorms unlike any witnessed on Earth. In turn, these clouds are overlain by ammonium hydrosulfide ice clouds (smelling of rotten eggs), that are themselves overlain by intricate bands of ammonia ice clouds. These

JUPITER'S CLOUDS imaged by Voyager Spacecraft

upper clouds make up the realm of intense turbulent storm activity that we see from Earth, including the famous Giant Red Spot.

Actual visits by humans to this gas giant would be problematic for another reason. Jupiter's internal structure gives rise to a massive magnetic field, and in turn results in large radiation belts around the planet. These radiation belts bask the region around the planet in a dose of radiation lethal to humans, so this domain is better explored by remote operated robotic spacecraft.

Saturn. As you continue to gaze into the night sky, you might notice another bright wanderer, but slightly less so as compared with Jupiter, but also traveling across the entire night sky. This wanderer was named after the Roman god of agriculture and harvest, perhaps as related to its brownish yellow color. With an orbital period of 29.46 years, this wanderer lies about 9.54 AU from the Sun or about 887,220,000 miles.

SATURN

Imagine you are in a desert region of Namibia, Africa. Stretching toward the horizon in all directions, the desert appears to consist of light-colored rock with rounded pebbles, and boulders strewn around the surface. Gazing up, you notice that the sky on this day is a hazy orange, which you attribute to dust and sand in the air. As your thirst wells up inside, you decide to go out in search for water and first decide to travel south. After a period of time, you come across a region where the light desert surface gives way to darker sands and vast sand dunes. You quickly find that walking in these sands is challenging. A hard crust on the surface of the sands generally supports your weight, but every now and then, your foot breaks through the crust and into what feels like damp sand. You also notice that flakes of this sand appear to be falling from the sky.

Because of the difficulty in walking, and the absence of water, you decide instead to walk north. After a long while, you see what appears to be a lake in the distance. When you arrive at the side of the lake, you get your water testing instruments out of your backpack to make sure the water is drinkable. To your surprise, your instruments tell you the water is actually liquid methane. With this realization, the environment around you begins to change in unpredictable ways. As a precaution, you put on your environmental suit that both keeps you warm and provides breathable air.

First, you feel as if you are getting lighter by the minute to the point where you feel as though you are about 10 percent what you weighed just minutes ago. Next, the temperature begins to plummet until reaching an unimaginably cold minus 290 Fahrenheit. Moreover, your instruments tell you that there is little or no oxygen in the atmosphere, but rather 95 percent nitrogen and five percent methane, and the atmospheric pressure is only 60 percent of what it was.

Next, it begins to rain from the orange hazy atmosphere that has not changed. Because of this rain, a dry river channel that you noticed on your journey begins to fill and flow. Your instruments again inform you that the flow in the river is liquid methane as well as ethane, but not liquid water. In desperation, you journey south again and return to your original location. You decide to search for water beneath the surface, hoping for an aquifer to provide you with this life-giving resource. You arrange for a large drill rig to be brought to your location, one that is capable of drilling to great depths beneath the surface. You drill for what would seem to be days, but the light never sets allowing you to take the time you need to complete your task. After drilling for a long while your rig reaches 30 miles beneath the surface. All of a sudden, a liquid begins flowing from the borehole that you have created. Your instruments tell you that the liquid is predominantly liquid water, but also ammonia and some methane. The liquid water seemingly warmed from below, flows as a liquid until the frigid temperatures surrounding you overcome it and it freezes across the surface.

You are on Saturn's largest moon Titan. This moon orbits Saturn every almost 16 hours but is tidally locked and so always shows the same face to its host planet. This is a planet made up largely of water ice, frozen at the surface, but thought to make up large lakes miles below that are heated by the core temperatures of this moon. "Volcanism" is thought to have dominated the surface features that are observed and that result in the desert planes made up of water ice or "rock", and strewn with pebbles and boulders also made of water ice.

The atmosphere of Titan consists mostly of nitrogen, and secondarily of methane. In the upper atmosphere, these molecules break down upon exposure to solar ultraviolet radiation and high-energy particles. The resulting "fragments" recombine into organic molecules, as well as nitrogen and oxygen. The heavier organic molecules fall to the surface and make up the sand and sand dunes encountered largely in the equatorial regions. The lighter organic compounds and gases that remain in the atmosphere result in a continuous haze that is orange in color.

Titan is one of the best candidates (Jupiter's moon Europa is another) of all the moons and planets in our Solar System to have established conditions potentially suitable for the development of life. While surface temperatures are frigid and water exists only as ice, far below the surface, water (mixed with salts and ammonia) is heated by core temperatures that may provide conditions where life could develop.

Titan's host planet Saturn is much like Jupiter, but smaller with a radius of about 72,000 miles, or about 9 Earth diameters. As with Jupiter, Saturn is made up of the primordial ingredients of a star but is too small to auto-ignite. If both Saturn and Jupiter were large enough, we would live in a triple star system. Moreover, Saturn's internal structure and atmosphere resemble Jupiter with metallic hydrogen surrounding the core, and an atmosphere dominated by hydrogen and helium with similar cloud layers including water ice clouds, overlain by ammonium hydrosulfide clouds, in turn overlain by ammonia ice clouds. These ammonia ice clouds of Saturn give this gas giant its yellow color.

Aside from Saturn's largest moon, what makes Saturn unique are the magnificent rings that surround the planet. That the rings are made up predominantly of water ice and some rocky material is not debated, but their formation is still contested. Some believe that the rings formed when a moon was ultimately ripped apart by tidal forces from the gravity of Saturn, or when the moon was destroyed as a result of a

SATURN'S MOON TITAN

comet or asteroid strike. Others believe that the rings are left over remnants from the original gaseous cloud from which Saturn originated during its period of accretion. What is less controversial is that this is one of the most beautiful and captivating of all the planets we can observe from Earth.

Uranus. The next wanderer was named after the Greek deity of the sky because of its resemblance to the color of the sky on Earth. Not constrained to the horizon, this wanderer moves slowly across the sky taking about the time of an average human lifetime to orbit the Sun. With an orbital period of 84 years, this wanderer lies about 19.18 AU from the Sun (1,783,740,000 miles).

Imagine you are walking toward the north pole on Earth sometime around twilight. While there is not much light, the area you are walking on appears to be ice as far into the distance as you can see. Above, the cloudy sky is a dull blue color and filled with clouds but no stars. You are in an environmental suit that provides warmth and supplied air and informs you about the nature of your surroundings. Your LCD display in your helmet tells you that temperatures have fallen to -371 Fahrenheit. Moreover, your display informs you that the atmosphere around you is now composed of hydrogen (82.5%), helium (15.2%) and methane (2.3%) gases with an overall air pressure about 1.3 times what it was moments ago.

URANUS

Despite the cold temperatures, the ice below you transforms to more of a liquid ice and you begin to fall below the surface, but before going under, you hop in a small rowboat. Next a deep greenish blue fog rolls in and envelopes all that you can see. You may think you will be all right, but then the winds pick up until they reach speeds as high as 510 miles per hour, and you wish you were somewhere else.

Unfortunately, you are not near the Earth's north pole, but on the ice giant planet Uranus, which has a diameter of 31,518 miles (about four Earths). The surface, if it can be distinguished from the thick atmosphere, while partially water ice, also consists of ammonia and methane ice that extends to the planet's core. Not a hard ice on which you could walk or land, it is rather a slushy ice through which the planet's gravity would cause you to sink.

Cloud layers above the surface include water clouds, overlain by ammonium hydrosulfide clouds, in turn overlain by ammonium or hydrogen sulfide (smells like rotten eggs if you are there). On the top is a layer of methane clouds that absorb red light wavelengths from the Sun and lead to the dull blue color of the atmosphere.

Notably, with an axial tilt of 98 degrees relative to its orbital plane around the Sun, Uranus experiences extreme seasons, each about 20 years long, with large black storm clouds appearing and disappearing as the atmosphere interacts with limited heat that reaches Uranus from the Sun.

Neptune. There is an eighth wanderer in the night sky, visible only with the assistance of binoculars or a telescope. This wanderer named after the Roman god of the sea was discovered through mathematics as predicted because of its influence on the orbit of the planet Uranus and only secondarily through direct observation. With an orbital period of 165 years, this wanderer lies about 30 AU from the Sun (2,790,000,000 miles).

Imagine yourself alone in a rowboat in the middle of the Pacific Ocean. It's twilight and there is little light to get a bearing on where you are. The dim light reveals a deep blue sky with high cirrus clouds suggesting bitter cold temperatures in the skies. You notice snowflakes falling from the sky that melt before reaching the surface.

While this might sound like an appealing scene, consider again as you discover that you are anywhere but in the Pacific. It begins with a realization that the "ocean" you are on is actually a slushy dense mixture of water, ammonia, and methane ice. Next, you realize that the atmosphere above you is composed not of breathable air but of hydrogen (80%), helium (19%), and lesser amounts of water and methane. The white cirrus clouds are composed of methane ice crystals. The snowflakes are composed of hydrocarbons that melt under the high pressures of the planet's atmosphere.

Suddenly the winds begin increasing and continue to do so far beyond any hurricane you may have previously experienced. In fact, if you could measure the wind speeds they would be upwards of 1,300 miles per hour and you would no longer be sitting comfortably in your boat. Moreover, you would be experiencing temperatures on average of -353 Fahrenheit.

NEPTUNE

As it turns out, you are not in the Pacific Ocean but on the deep blue planet Neptune, an ice planet with a diameter about four times that of the Earth and accompanied by a modest ring system and fourteen moons, including the largest Triton. Neptune is tilted about 28 degrees on its axis so that it experiences seasons just as Earth and Mars do, but each season lasts 41 years, whereas a day on Neptune is only 16 hours as a result of its quick rotational speed. Neptune's weather is characterized by the highest wind speeds and among the coldest temperatures in our Solar System. Moreover, as a result of the planet's composition and extreme pressures, it is theorized that methane in the depths of the Neptunian ocean crystallizes and a hailstorm of diamonds pour down toward the planet's core.

Pluto. A final wanderer lies in the distant expanse of space and is perpetually in a state of relative darkness. Because of its dark existence, this dwarf planet was named after the Roman god of the underworld.

Imagine that you are trekking in dim light across Antarctica in search of the South Pole. All around you are plains of water ice as you are passing over a frozen continent. In the distance, you see mountain ranges seemingly made entirely of water ice rising from the plains. You are in the region where in August 2010 NASA satellites recorded the lowest ever ambient temperature on Earth of -136 Fahrenheit.

Now imagine that temperatures begin to decline and the air pressure begins to fall. At some point, your instruments tell you that temperatures have fallen to -380 Fahrenheit and the atmospheric pressure is now one one-millionth of what it was. Because the temperature has fallen below -346 Fahrenheit (freezing temperature of nitrogen), a layer of nitrogen ice has formed over the water ice. The plains around you, no longer just white, now also reveal areas varying in color from charcoal black to orange. Furthermore, all that remains of the atmosphere is a tenuous presence of nitrogen, methane and carbon monoxide all in equilibrium with the ice around you.

You are now not in Antarctica, but rather on the dwarf planet Pluto during a Plutonian winter that is not dramatically different from a Plutonian summer except for there being light. Pluto is only a little larger than half the diameter of the Moon that orbits Earth, and it follows an eccentric orbit taking 248 years to complete a revolution (Plutonian year). Moreover, this planet lies at a staggering distance from the Sun varying from 30 AU (2.8 billion miles) to 49 AU (4.6 billion miles).

PLUTO

A day on Pluto lasts about 6.4 Earth days, though the term "day" is perhaps overstated. Because of its distance from the Sun, light itself takes around 5.5 hours to arrive. When it does arrive, the amount of light is only a fraction of a percent of that which you would experience on Earth. Moreover, Pluto is tilted 120 degrees from its orbital plane around the Sun, such that it experiences extreme seasons where one-quarter of the planet is either in complete darkness or continual daylight for an entire solstice season (upwards of 62 years).

Perhaps most interesting is that this dwarf planet has five known moons including the largest Charon. On the northern hemisphere of Charon is a reddish-brown region. The composition of this region has been identified as tholins, which are organic compounds formed from carbon dioxide, methane, ethane, and nitrogen exposed to ultraviolet light and cosmic rays from the Sun. When exposed to water, tholins form prebiotic chemistry with implications for the development of extraterrestrial life.

The Nearest Stars

Now that we've gotten to know our nearest neighbors, consider our neighbors further out – the stars. Look up again at the night sky and on the best of nights, you will see upwards of 4,500 stars (up to magnitude of 6.5) all with varying brightness's and if you look closely, hints of color. All the stars that you see, not including the Milky Way or the notable exceptions such as visible nebula and galaxies visible to the unaided eye, represent our nearest neighbors outside of our Solar System.

However, as with the planets, it is necessary to consider how to measure distance when considering the stars because the distances involved are enormous. At the scale of interstellar space, distances from Earth to the stars are measured in how long it takes light to travel from the star to the Earth where we observe it. The speed of light has been calculated to be 186,282 miles per second, such that light would travel on the order of 6,000,000,000,000 (6 trillion) miles in a year. So, the adopted unit of measurement in interstellar space is a light-year, or the distance light travels in a year.

The closest of our nearest neighbor stars is actually a triple star system consisting of Alpha Centauri, Beta Centauri, and the smaller red star Proxima Centauri. This star system lies in the constellation Centaurus that is visible from the southern hemisphere. This star system is estimated to lie 4.37 light-years (26,220,000,000,000 miles) from the Sun.

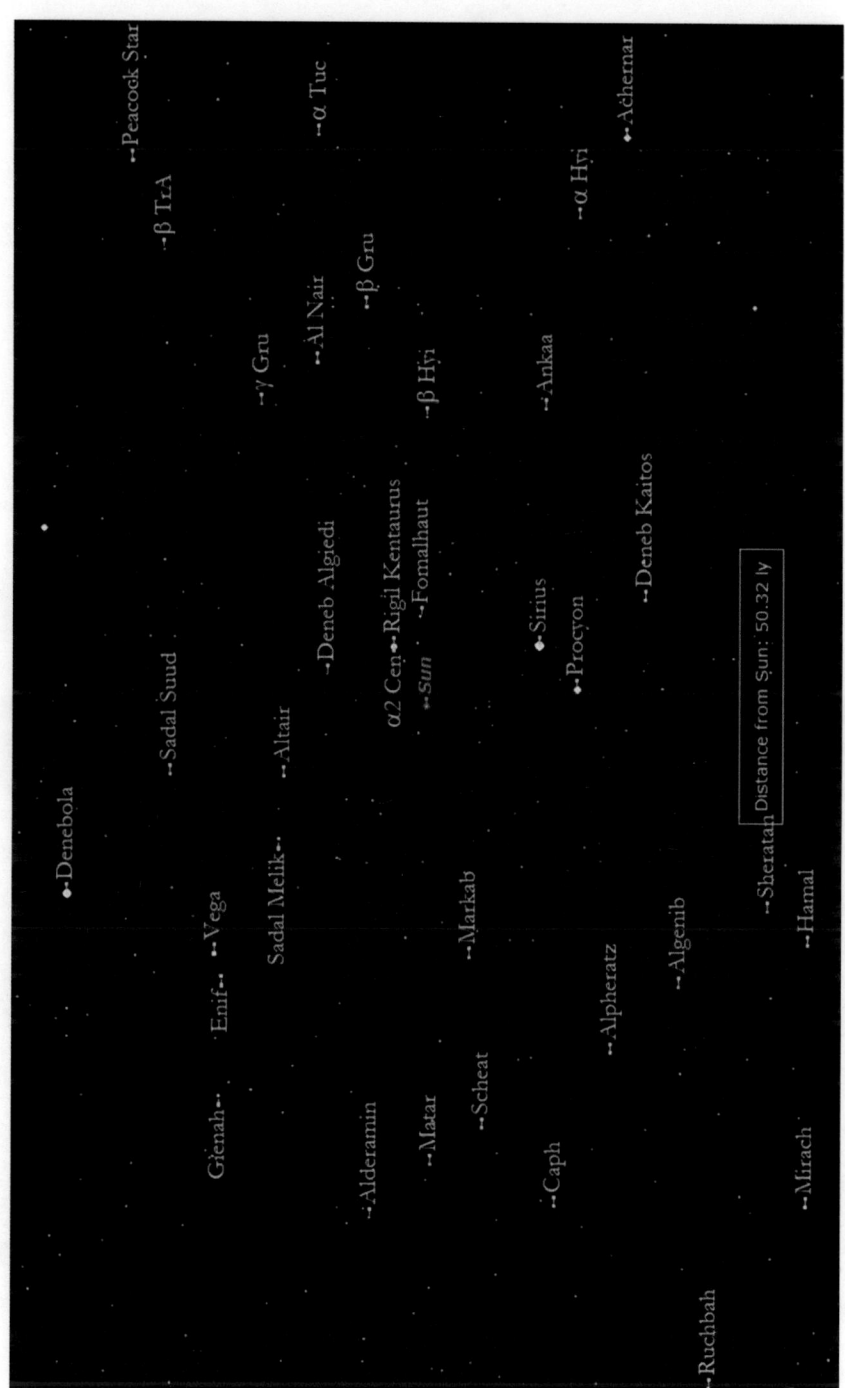

LOCAL STARS in our galactic neighborhood

ALPHA CENTAURI A AND B imaged by Hubble Telescope

The distances even to the nearest of stars are difficult to comprehend, but by scaling down, we can begin to get a grip on exactly how far away they are. Consider Phoenix, Arizona where many of the streets are built on a grid where the distance is exactly one mile from one intersection to another. Now conduct a thought experiment in which the Sun lies at the center of one intersection. The Sun on this scale will be roughly the size of the yellow signal in the traffic light. On this same scale, Pluto would lie approximately one mile away at the next intersection and would be represented by no more than a speck of dust. On this same scale, the Alpha Centauri star system would lie around 7,000 miles away somewhere in the vicinity of Sydney, Australia.

Another one of our nearest neighbors and the brightest star in the night sky is Sirius (also known as the Dog Star) that is found in the constellation Canis Major. Sirius can be seen below and to the left of the constellation Orion in the northern hemisphere. Sirius, an estimated 8.7 light-years from our Sun, turns out to be a binary star system consisting of Sirius A and Sirius B.

Alpha Centauri and Sirius are only two-star systems among thousands that lie in our vicinity and make up our nearest neighbors. However, understanding which stars are in our vicinity is to only partially begin to grasp where we live. As it turns out, all the stars visible

SIRIUS – an artist's rendition

to us in the night sky, again not including visible galaxies, lie within a larger structure, which is our Milky Way Galaxy. Based on research of everything we can see from Earth (much of the galaxy is behind the central bulge of the galaxy and is invisible to us on Earth), the Milky Way Galaxy is understood to fit in the category of a barred spiral galaxy. As it turns out, all the stars visible in the night sky from Earth represent those stars in the minor Orion Arm or other adjacent arms, within the larger structure that consists of two major arms including the Scutum-Centaurus and the Perseus Arms. When we gaze in wonder at the Milky Way in the night sky, what we are seeing are the collective stars and gases making up the other arms and central bulge of our galaxy, except that they are much further away making them seem more like a cloud of "spilled milk" than individual stars.

As it turns out, our Milky Way Galaxy is estimated to be on the order of 100,000 to 180,000 light-years in diameter, with our Solar System lying approximately 26,000 light-years from the galactic center within the Orion arm. The Milky Way Galaxy is estimated to contain around 250 billion (plus or minus 150 billion) stars.

But the description of our cosmic neighborhood doesn't stop here. While we reside in a star system within the Milky Way Galaxy, our galaxy itself is part of a local group of 54 galaxies, most prominent of

THE MILKY WAY consists of countless distant stars making up the other arms of our barred spiral galaxy, while the stars that fill our night sky make up the arm of our galaxy in which we reside.

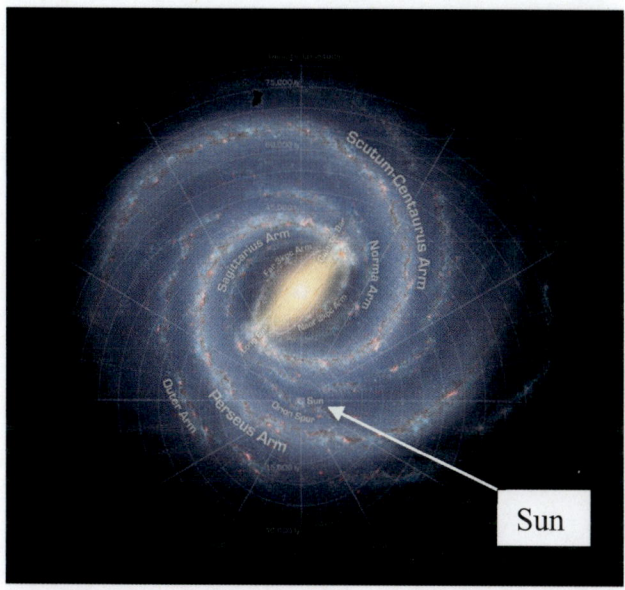

OUR LOCATION within the Milky Way Galaxy

which is the Andromeda Galaxy, estimated to lie 2.54 million light-years (15,240,000,000,000,000,000 or 15.24 quintillion miles) from the Sun. The local group in which our galaxy lies is an estimated 10 million light-years across. The local group is itself part of a larger supercluster of galaxies known as the Virgo Supercluster.

While most galaxies we observe are moving away from ours (expanding universe), local gravitational forces are causing the Andromeda Galaxy and the Milky Way Galaxy to move towards each other with Andromeda estimated to be moving towards us as 68 miles per second. In an estimated 4 billion years, the two galaxies are projected to collide, whereas our Sun (and therefore Earth) are expected to last another 5 billion years. Galaxies can be cannibals, and a larger galaxy may effectively consume a smaller galaxy.

The Andromeda Galaxy is thought to contain up to 1 trillion stars and to have a diameter on the order of 220,000 light-years making it more than twice the diameter of the Milky Way Galaxy! The Whirlpool Galaxy is an example of the interaction between two galaxies.

And so it goes as you continue to move out beyond our local group of galaxies. Between 2003 and 2004, the Hubble Telescope was pointed at an otherwise dark blank portion of space covering one-tenth

ANDROMEDA GALAXY

WHIRLPOOL GALAXY

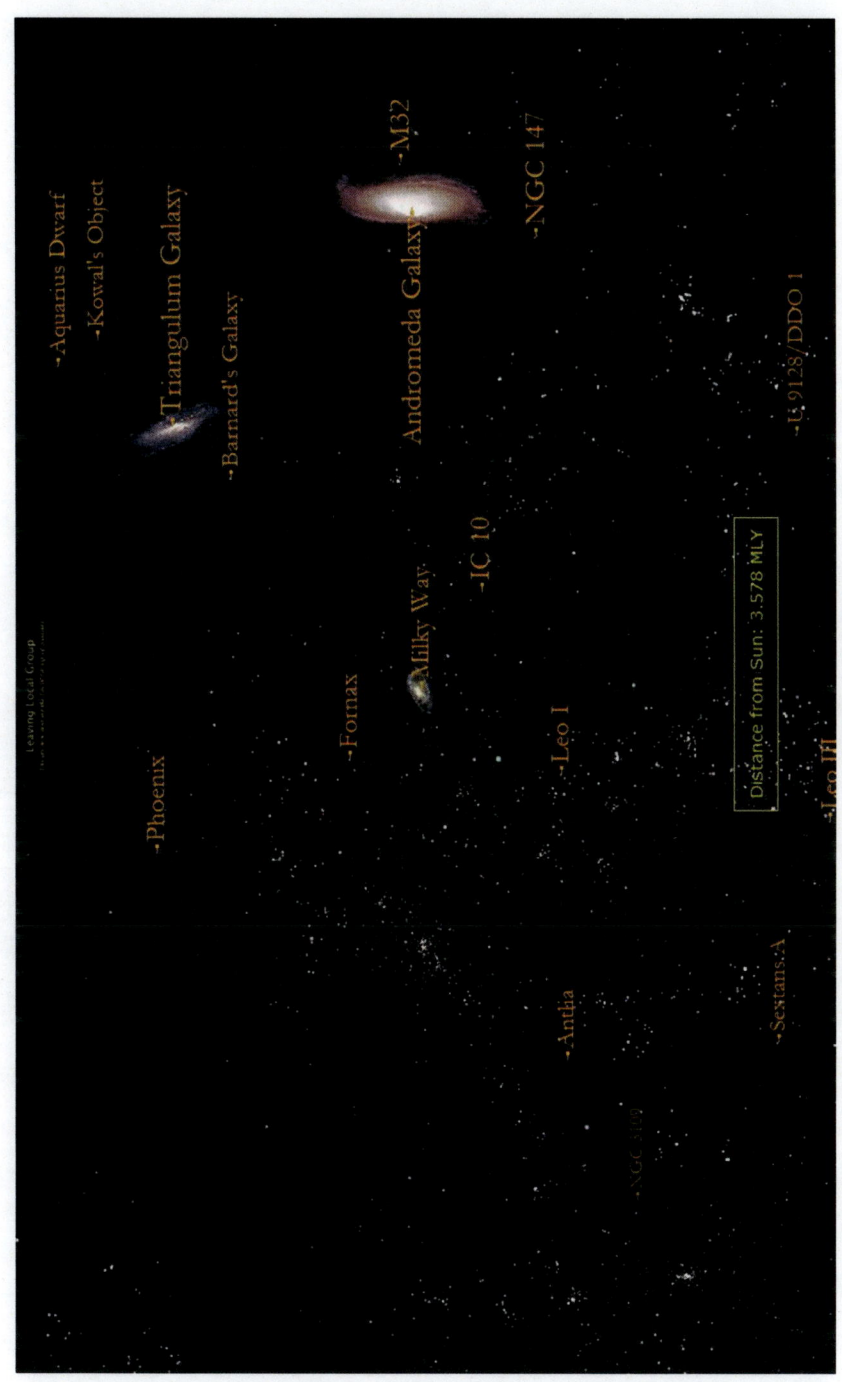

OUR NEIGHBORHOOD (local group) of galaxies

of one-millionth of the night sky. The telescope collected the image over a period of several months. The result was an image known as the Hubble Ultra Deep Field, which was filled with an estimated 10,000 distant galaxies. Extrapolating this across the entire sky, astronomers estimate that there could be upwards of hundreds of billions of galaxies in the visible Universe.

Revisiting Our Cosmic Address

So, in consideration of our position within our Solar System, of our nearby neighbor stars, the galaxy in which we reside, and the sea of galaxies that make up our universe, we revisit the question – where do we live? With this new perspective, perhaps a more appropriate description of our cosmic address would be as shown below.

Our Cosmic Address:
Earth, Third Planet in Sun Solar System
Orion Arm/Spur (26,000 light-years from galactic center)
Milky Way Galaxy
Andromeda Local Group, Virgo Supercluster, Visible Universe

HUBBLE ULTRA DEEP FIELD

CHAPTER 2

TIME MACHINES

"People like us, who believe in physics, know that the distinction between past, present, and future is only a stubbornly persistent illusion."
— Albert Einstein

"When you are courting a nice girl, an hour seems like a second. When you sit on a red-hot cinder a second seems like an hour. That's relativity."
— Albert Einstein

The meeting of two eternities, the past and future....is precisely the present moment.
— Henry David Thoreau

What Time Is It?

If someone were to ask you what time it is, you might respond by looking at your watch or calendar in order to share an accurate assessment of the time. For example, one might provide a detailed evaluation of the time by responding that it is 9:30 AM, Saturday the fifth of May in the year 2018. And while such a response might be received as an accurate presentation of the current time, what is the origin of our means of tracking time, and does this have relevance in the cosmic perspective?

Archeological evidence suggests that methods of timekeeping date back as far as approximately 10,000 BC, corresponding with the Neolithic period, a time characterized by growth in early human technology. Fundamentally, timekeeping has had its origins in our experience on Earth of daylight versus night (day), the time it takes for the Moon to transition from a new moon, through its repeated phases, and returning to a new moon (month), and the time it takes for seasons to cycle corresponding to the time it takes the Earth to revolve 360 degrees around the Sun (the year).

The word "day" has its origin in the old English word "daeg" (itself having Germanic origins) that originally referred to the period of daylight but was later adopted to refer to the 24-hour period inclusive of both day and night. Our current calendar system based on a seven-day week originated from the Babylonians, who in turn based their calendar on a 21st century B.C. Sumerian calendar. This seven-day calendar was adopted by the Roman Empire during the 1st and 3rd centuries anno domini (A.D., from Medieval Latin and meaning the year of the Lord). As for the names ascribed to each day of the week, these had their origin in Greek and Roman mythology with the exception of Sunday and Monday that have their origin in Germanic or Norse mythology (see table on next page).

The first formal calendars in Europe date back to the Bronze Age, which was a historical period characterized by the use of bronze, and corresponding to the time between 3200 and 600 before Christ (B.C., both A.D. and B.C. had their origin in the Julian and Gregorian calendars as a means for tracking the number of years). In 45 B.C., Julius Caesar reformed the Roman calendar, based on the Julian calendar, through the use of an algorithm and the addition of a leap day every four years, rather than depending on the timeframe for the moon to move through its phases. In October 1582, Pope Gregory XIII introduced the Gregorian calendar, a slight revision of the Julian calendar, and remains the standard calendar worldwide in the present day for secular purposes.

As we consider how we track time and the origin of our system for doing so, we gain a wider perspective when we consider how we might track time elsewhere, such as on other planets in our Solar System.

Consider Mercury that orbits the Sun in 87.969 Earth days (one Mercurial year), but rotates only every 58.6467 Earth days as related to this planet being gravitationally locked with the Sun, which is technically a Mercurial day. As a result, it takes a particular point on the equator of Mercury exactly two full orbits of the Sun to return to the same position relative to the Sun, during which time Mercury rotates on its axis three times. This time, which calculates to be 175.938 Earth days could be viewed as a Mercurial day such that a particular

Day of Week	Associated Celestial Body	A Deity from Ancient Religion and Myth	Latin Origin of Name	Day of Week in Spanish
Sunday	Sun	Sol, the sun god	diēs Sōlis (day of the Sun)	Domingo
Monday	Moon	Luna, female complement of the Sun	diēs Lūnae (day of the Moon)	Lunes
Tuesday	Mars	Mars was the god of war	diēs Martis, (day of Mars)	Martes
Wednesday	Mercury	God of multiple purposes such as financial gain and commerce	diēs Mercuriī, (day of Mercury)	Miércoles
Thursday	Jupiter	Also known as Jove gen. Iovis, god of the sky, thunder, and king of the gods	diēs Iōvis (day of Jupiter)	Jueves
Friday	Venus	Roman goddess of love and beauty	diēs Veneris (day of Venus)	Viernes
Saturday	Saturn	God of multiple purposes such as wealth and agriculture, and later of time	diēs Saturnī (day of Saturn)	Sábado

ORIGIN OF DAYS OF WEEK

point would experience the complete cycle of noon, midnight and back to noon. So, the length of a Mercurial day is exactly double the length of a Mercurial year. Moreover, there is almost no tilt to the axis of Mercury and consequently there are no seasons. However, within the course of a day on Mercury, and because this planet has no appreciable atmosphere, Mercurial "seasons" could be established based on the extreme temperature variations from "season" to "season", with each "season" lasting on the order of 44 Earth days, or one quarter of a Mercurial day. So "summer" temperatures (mid-day) would be as much as 700 Kelvin (800 Fahrenheit), "winter" temperatures averaging 100 Kelvin (-280 Fahrenheit), and "mid-season" temperatures somewhere in between. As such, a possible Mercurial calendar is presented below.

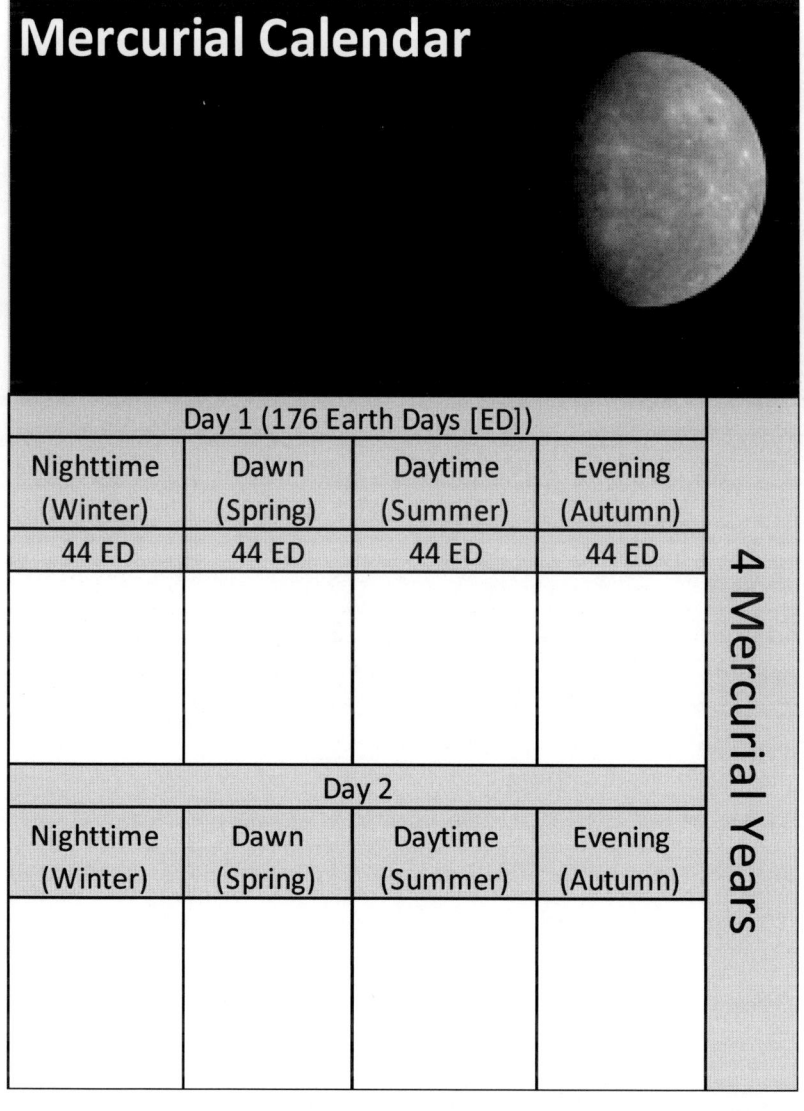

Mercurial Calendar

Day 1 (176 Earth Days [ED])				
Nighttime (Winter)	Dawn (Spring)	Daytime (Summer)	Evening (Autumn)	
44 ED	44 ED	44 ED	44 ED	4 Mercurial Years
Day 2				
Nighttime (Winter)	Dawn (Spring)	Daytime (Summer)	Evening (Autumn)	

Now consider Mars, which orbits the Sun every approximately 687 Earth days, and rotates on its axis every 24 hours 39 minutes and 35.244 seconds. In addition, Mars is tilted about 25 degrees relative to its orbital plane around the Sun. As a result, a Martian calendar would resemble Earth's with a day being about the same, and months and years being almost double in length. Similarly, seasons on Mars would also be about double the length as compared with Earth. As such, a Martian calendar might look as shown below.

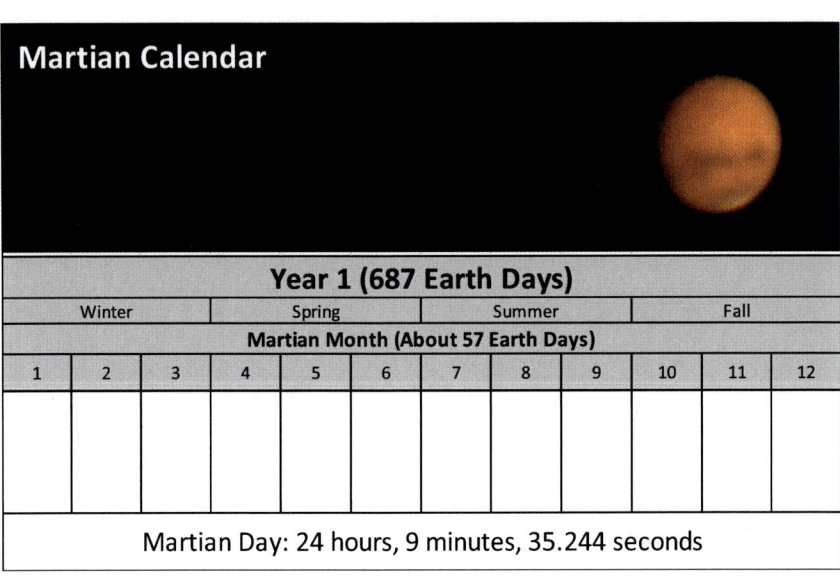

Martian Calendar											
Year 1 (687 Earth Days)											
Winter			Spring			Summer			Fall		
Martian Month (About 57 Earth Days)											
1	2	3	4	5	6	7	8	9	10	11	12
Martian Day: 24 hours, 9 minutes, 35.244 seconds											

Lastly, consider the dwarf planet Pluto. Pluto orbits the Sun approximately every 248 Earth years (a Plutonian Year), and rotates on its axis approximately every 6.4 Earth days (about 153 Earth hours; a Plutonian Day). In addition, Pluto's axis is angled 120 degrees relative to the orbital plane around the Sun and effectively orbits on its side. Pluto experiences extreme seasons where during the solstices, when the Sun reaches its most northerly or southerly point on the planet, almost a quarter of the planet is in constant sunlight and another quarter in constant darkness. A Plutonian calendar might look as shown on the next page.

So, how we track time depends largely on where we live, and in particular where we live in the Solar System. And the question of what time it is becomes more challenging from other perspectives within our Solar System.

Plutonian Calendar

Plutonian Year (248 Earth Years)											
Plutonian Seasons/Months (each month around 21 Earth Years)											
Winter			Spring			Summer			Autumn		
1	2	3	4	5	6	7	8	9	10	11	12
14,144 Plutonian days in a Plutonian year, each lasting 6.4 Earth days											

The Night Sky and Time

Now look up again at the night sky and ask the question again, what time is it? What you will see with the best of eyes and in the darkest of night skies are approximately 4,500 stars in either hemisphere. What you learn about time in the night sky may surprise you, but first you need to understand the distances involved to the stars.

Astronomers use several methods to estimate the distance to objects in the night sky. One of those methods is based on understanding the brightness of a star in the night sky, its apparent magnitude, and the brightness that star would be, its absolute magnitude, if it were at an arbitrary distance away which is taken to be 32.6 light-years or 10 parsecs. One parsec is equivalent to 3.26 light-years and represents the distance from the Earth at which point the star would have a parallax angle equal to one second of arc. Note that with magnitude, the lower the value, or more negative, the brighter the object in the night sky. The equation for this estimation is as follows:

$$M = m + 5 - 5\log(r) \quad \text{(Equation 1)}$$

or alternatively,

$$r = 10^{((m-M)+5)/5} \quad \text{(Equation 2)}$$

Where M = absolute magnitude

m = apparent magnitude

r = distance in parsecs

The challenge in performing this calculation is knowing a particular star's absolute magnitude, or the brightness it would have in the night sky if it were at the arbitrary distance of 10 parsecs. The absolute magnitude has been worked out for many of the stars in the night sky and for the purposes of considering the distances to stars, these estimations will be applied.

Sirius. First, consider the brightest star in the night sky, Sirius, which lies in the constellation Canis Major (just below and to the left of the constellation Orion in the winter night sky). Sirius is a binary star system with Sirius A, the brighter of the two, having an apparent magnitude of -1.46. Astronomers have worked out, based on characteristics of the star such as its color, that is brightness at 10 parsecs (i.e., its absolute magnitude) would be +1.42, which is dimmer than its apparent magnitude in the night sky and suggests the star is closer than 10 parsecs. Based on Equation 2, this works out such that Sirius would be about 8.6 light-years from Earth.

Now consider that in order to reach your eyes, the light from Sirius needed to first travel the distance between Sirius and Earth. From above, this distance is about 8.6 light-years or 51,600,000,000,000 miles. At the speed of light, which is 186,282 miles per second, it would take light itself 8.6 years to traverse this distance. As such, the light reaching your eyes would tell the story of Sirius as it was 8.6 years into the past. As of the date of this writing, 2018, the light from Sirius that is now reaching our eyes began traveling across interstellar space around the time that the 21st Winter Olympics were opening in Vancouver, Canada. Conversely, if there were a visible planet adjacent to Sirius, and our telescopes were powerful enough, we could peer into the past and witness events unfolding on that distant planet around the same time as the Vancouver Winter Olympics.

Mizar. Next, look to the star Mizar, which is the second star from the end of the handle in the Big Dipper. Mizar, early thought to be a binary star system, turned out to have as many as six stars. With an apparent magnitude of +2.23 and an absolute magnitude of +0.33, this star can be estimated to be approximately 78 light-years distant. By the same line of thinking as was presented for Sirius, the light from the Mizar star system began traveling through interstellar space and tells the story of that star system as it was 78 years ago, or around 1940 Earth time. We are seeing events now that occurred in the Mizar star system at the time Hitler was continuing to invade Europe, and we were in the midst of the Great Depression.

MIZAR SIX-STAR SYSTEM in Ursa Major

North Star (Polaris). Now consider Polaris in the constellation of Ursa Minor (the Little Dipper). Somewhat dimmer, this star with an apparent magnitude of +1.98 and an absolute magnitude of -3.6 turns out to be part of a triple star system. Based on its brightness, Polaris is estimated to be about 426 light-years distant. The light from Polaris began its journey across interstellar space in the Earth year 1592, shortly before Galileo's 1609 first observations of Jupiter through a telescope that revolutionized our understanding of the Universe. So again, if our telescope were good enough, we could see events unfolding on a nearby planet to Polaris that corresponded in time to events occurring here on Earth around the time that Galileo was first looking and making his revolutionary observations.

Antares. Looking now to the southern summer sky, if you are in the northern hemisphere, you will observe the constellation Scorpio. Within this constellation, you will see the bright red variable star

Antares. Antares is a dying star experiencing the final stages of the stellar lifecycle, and consequently, its apparent magnitude as seen from Earth varies from +0.6 to +1.6, with an average of +1.1. With an absolute magnitude of about -5.28, the distance to Antares works out to be about 615 light-years. The light from Antares began its journey across interstellar space in the Earth year 1503, shortly after Columbus braved the unknown Atlantic Ocean and first landed on an Island he called San Salvador in the Bahamas. Again, if our telescopes were good enough, we could see events unfolding on a nearby planet to Antares, that corresponded in time to events occurring here on Earth in 1503. This assumes the dying star has not also consumed any nearby life-bearing planet.

The Great Orion Nebula. Now look to the middle of the sword in the constellation of Orion. This is the Orion Nebula, the closest region to our Solar System of massive star formation in our galaxy. The distance to the Orion Nebula has been estimated to be 1,344 light-years.

ORION NEBULA

The light from this region began its journey through interstellar space around 674 A.D., roughly corresponding to events on Earth such as:

- Muhammad's dictation of the Koran (625 A.D.),
- The Vikings begin invasions of Ireland (620 A.D.),
- Muslim capture of Jerusalem (638 A.D.), and
- Japanese withdrawal from Korea (663 A.D.).

Andromeda Galaxy. Looking now at the constellation Andromeda, a faint fuzzy object is visible to the naked eye. This object is the Andromeda Galaxy, which is significantly larger than our own Milky Way Galaxy, and is estimated to lie 2.43 million light-years distant. Light from the Andromeda Galaxy began its journey through space before scientists estimate that homo erectus (meaning "upright man") first walked the Earth as far back as 1.8 million years ago. If intelligent life existed 2.43 million years ago in the Andromeda Galaxy, and if these lifeforms had telescopes of appropriate sophistication to observe the Earth at that time, the light from Earth would just be reaching them, and as far as we understand, they would find no evidence of intelligent life and might look elsewhere in their search.

ANDROMEDA GALAXY at 2.43 million light years

Messier 61. Now looking through a telescope toward the constellation Virgo, an intermediate barred spiral galaxy cataloged as Messier 61 (M61) can be found. This is a galaxy where it has been observed that star formation is occurring at an exceptionally high rate as compared with other galaxies. Astronomers estimate this galaxy to be on the order of 52.5 million light-years distant. The light from this

MESSIER 61 GALAXY

galaxy began its journey through interstellar space about 12 million years after we understand the last dinosaurs walked the Earth (i.e., 65 million years ago at the end of the Mesozoic Era).

Time Machine

As is evident, the night sky is a time machine. It is a window into the recent and distant past where the better your eyes, or your instrumentation, the further into the past you can see. When you ponder the answer to the question of what time it is in the cosmic arena, your answer seems to depend on where you are looking. However, for all the objects in the night sky, regardless of how long it takes the light from the objects to reach Earth, there are events occurring simultaneously across the Universe in what we could refer to as "now", which could be your initial revised answer to what time it is in the cosmic arena.

Einstein And Time

We have considered the current time based on our system of time measurement (i.e., the calendar), and we have considered the implications of time when viewing the night sky, but now we consider time through examination of the theory of relativity proposed by Albert Einstein as published in 1905 (Special Theory of Relativity concerning spacetime) and 1916 (General Theory of Relativity concerning gravity). According to this theory, space consists not only of the three physical dimensions but also includes the dimension of time, hence the concept of spacetime. Time, and in particular the rate

ALBERT EINSTEIN

at which time passes, as it turns out, is not constant as we have come to believe from our everyday experience. Einstein's theory of relativity offered several revelations with regard to time, a few of which are described here and based on similar presentations by Dr. Brian Green as host of the NOVA production "Fabric of the Cosmos".

If you are out walking and find yourself moving towards someone sitting on a park bench straight in front of you, relativity implies that time is moving more slowly for you (the person moving) than the person sitting on the bench. Conversely, the rate at which time is advancing is imperceptibly faster relative to the person who is moving. This is the notion of time dilation that is suggested by special relativity.

The difference in the rate of time is not noticeable in our everyday experience, but the variance in time has been tested and shown to be real. For the last several decades, atomic clocks, which are based on the measurement of microwave emissions as electrons change state in cesium-133 atoms on a regular and predictable schedule, have been aboard GPS satellites and have measured the rate of time as compared with similar atomic clocks on Earth. The clocks on the satellites have recorded a slower rate of time advancement which validates Einstein's special relativity.

The concept of time dilation is illustrated in the graph below where an object in motion moves through time more slowly relative to an object that is not moving.

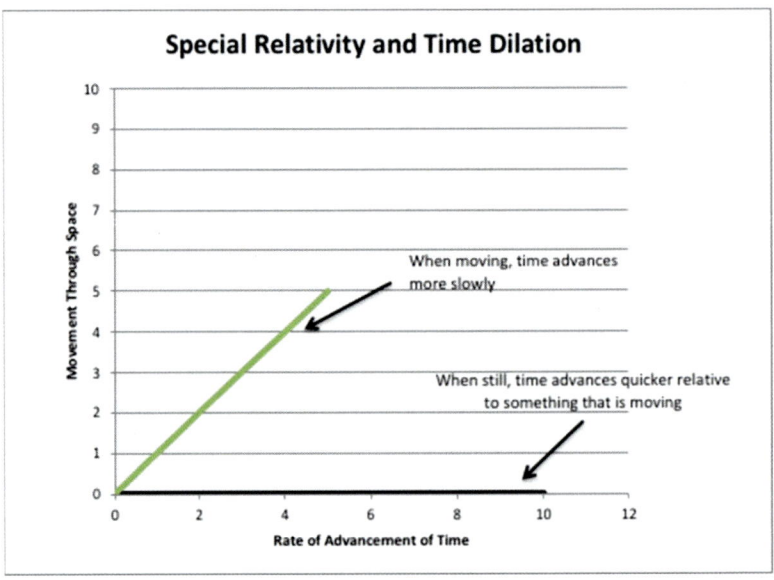

Taking the concept of time dilation further, consider the motions of two beings on opposite sides of the Universe. Take one, the same person sitting on the bench with no movement (ignoring the rotation of the Earth, or of the movement of our Solar System through interstellar space), and another life form present on a planet at the same moment on the opposite end of the Universe say 14 billion light-years distant that is similarly not moving. Since both are immobile relative to each other, they experience the same time (the same "now") and the same rate of advancement of time. But if the life form on the distant planet begins walking directly away from the earthling, they begin to experience time advancing at different rates.

The life form walking away experiences time advancing more slowly relative to the earthling, and over the vast distance involved and in consideration of a four-dimensional spacetime Universe, the time dilation becomes significantly pronounced and the distant life form is then experiencing a time that corresponds to our past, even our distant past. Conversely, if the distant life form turns around and begins walking toward the earthling, this effect of time dilation is reversed with the distant life form then sharing a time that corresponds to our future, or our distant future. In this sense, and as consistent with special relativity, it is implied that all times coexist and can be experienced depending on where you are in the Universe and how you are moving. Stated another way, the past, present, and future could be described as coexisting "simultaneously", though it raises serious questions about events of the past and whether they could be affected through time travel.

Einstein's theory of general relativity has similar implications for the advancement of time. According to this theory, gravity will have the same effect as motion with regard to changing the rate at which time advances. The greater the mass of the object, the greater the gravity (i.e., warping of spacetime) and the slower the rate at which time advances. So, in 1953, when New Zealand mountaineer Edmund Hillary and Tenzing Norgay of Nepal made the first confirmed ascent of Mount Everest, they were experiencing time at an imperceptibly faster rate as compared with those residing elsewhere at sea level. Along the same lines, if you were to travel to a black hole where the gravity and warping of spacetime is most extreme, and assuming you weren't killed in the experience, then time would theoretically slow down significantly for you as compared with others back on Earth and in this sense time travel into the future would be possible in that upon your return to Earth, others could be decades older whereas you may have aged only a few hours by comparison.

So What Time Is It?

In consideration of our system of time measurement, our observations of the night sky, and in view of Einstein's thought experiments that led to his revelations about spacetime, the answer to the question – what time is it? – is no longer clear. It seems the answer to what time it is depends our perspective. Depending on which planet we are residing, and depending on how we are moving and where we are in spacetime, we might describe the time very differently. As consistent with the theory of relativity we might conclude that there is no answer to the question of what time it is, only our individual experience of time.

CHAPTER 3

COLD DARK SPACE

"The only true wisdom is knowing you know nothing."
— Socrates
"I believe the more I study science, the more I believe in God."

— Sir Isaac Newton

GALAXY M81 at 11.8 million light years distant in Ursa Major. One of the billions of galaxies in our Universe whose destiny may be to fade into darkness.

Cold Dark Space – The Future of the Universe?

Close your eyes for a moment and imagine yourself lost in the vast ocean of space absent any light whatsoever in all directions, and temperatures approaching -459 Fahrenheit (absolute zero). Moreover, imagine that these conditions last for an eternity devoid of light and life. This is a pretty gloomy and lonely picture, and yet could be what is in store for our Universe.

With the expansion of our Universe, astronomers ask whether the expansion will continue forever, or if it will under the force of gravity ultimately contract and return to the state of singularity (all matter and spacetime in a single point of infinite density) that is understood to have pre-existed the Big Bang. Evidence points to eternal expansion as related to so-called dark energy (an unknown energy hypothesized to permeate all of space) that causes the expansion of the Universe to accelerate.

So, if the Universe expands into eternity, it will ultimately burn out as stars run out of fuel and galaxies fade into eternal darkness. The Universe will be an immense cold dark space devoid of light and life. So, if the fate of the Universe is to become a cold dark place, we who reside on the speck of dust (on the cosmic scale) we call Earth ask again those questions that have been asked since the first humans walked the Earth:

Why are we here?
What is the meaning of life?
Is there a greater purpose for our lives?

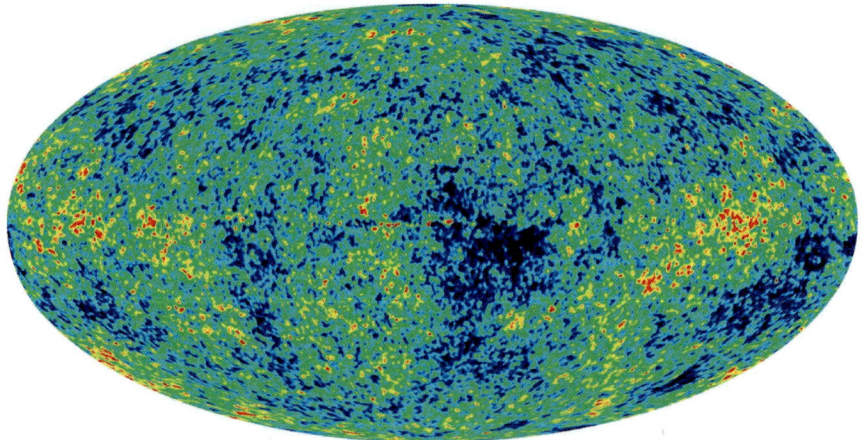

MICROWAVE PHOTOGRAPH OF THE UNIVERSE revealing 13.8-billion-year-old temperature variations representing the early seeds of galaxies.

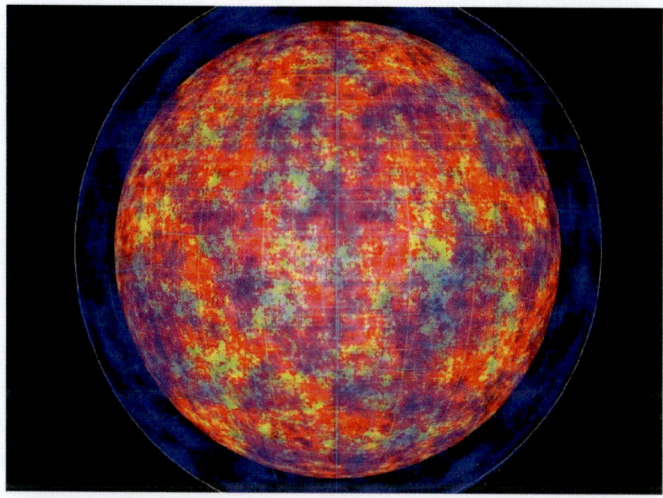

THE AGE OF THE UNIVERSE is estimated to be 13.8 billion years, whereas the edge of the visible Universe is estimated to extend to 46 billion light years from Earth as it has continued to expand. Beyond this, matter is theorized to be expanding at a rate greater than the speed of light as a result of the metric expansion of spacetime in accordance with general relativity such that the light from objects further out will never reach the earth.

The Limits of The Cosmos – The History of Our Understanding

Ever since humans have looked up into the night sky, we have been speculating as to how big the Universe really is. In the distant past, there were those who looked up and postulated that the night sky was a solid black barrier strewn with holes through which we could see glimpses of heaven beyond. Others thought perhaps that the lights in the night sky represented campfires from people far away. Eventually we got it right and discovered that the lights in the night sky were in fact distant stars like our own Sun, but very far away. Beginning with many ancient civilizations, and maintained by Christian theologians as late as the 17th century A.D., it was thought and enforced that all these stars revolved around the Earth, and that the Earth was the focal point of creation and the center of everything.

In 1609 Galileo Galilei first pointed a telescope toward the night sky and began to unravel the paradigm of understanding of the heavens suggesting that the Sun, not the Earth, was the center. This so-called heliocentric theory was deemed heretical and in conflict with the Holy Scriptures by the Roman Inquisition. Galileo was forced to recant and ultimately sentenced to imprisonment, commuted to house arrest. What Galileo had first observed were the four primary moons (Io, Europa, Ganymede and Callisto) around Jupiter that were revolving around that planet and not the Earth, the first clue that the Earth was not at the center.

Ultimately with the advancement of thinking and mathematics by Nicolaus Copernicus and Johannes Kepler, humans worked out that the planets actually revolve around the Sun, with the Earth being just another, not unordinary, planet. Ultimately humans worked out that not only was the Earth not the center of everything but that our Solar

GALILEO GALILEI PORTRAIT

JUPITER AND THREE OF ITS MOONS

System (centered on our Sun) was actually part of a vast galaxy that had its own central point.

As telescopes advanced and observations of the night sky increased, other objects were observed in the heavens. These objects were fuzzy and not points of light as with the stars. While many of these objects turned out to be gaseous nebula within our Milky Way Galaxy, it was in the 1920s that Edwin Hubble worked out the distance to many of these objects, which turned out to be distant galaxies far outside of our own Milky Way Galaxy. Humans began to grasp that they inhabited at best a spec of cosmic dust in a vast Universe much larger than previously understood.

EDWIN HUBBLE

In the 20th century, humans put the Hubble Telescope (named in honor of Edwin Hubble) in space and have since gained further knowledge into the depths of space. The Hubble Telescope imaged the most distant object ever observed by humans – the so-called GN-z11 Galaxy estimated to be 13.4 billion light-years distant (at least at the time the light from this object began traveling through intergalactic space towards the Earth). So, the more time that goes by and the better our technologies become, we seem to continue to discover that there is more to the Cosmos than we had previously understood or had even thought to consider. To this end, it is entirely plausible that there is far

SOMBRERO GALAXY M104 at 31.1 million light years distant in the constellation Virgo

PINWHEEL GALAXY M33 at 2.4 million light years distant in the constellation Triangulum

more that we don't know about the Cosmos than our current collective knowledge allows us to understand. It has been speculated that there are multiverses (i.e., an infinite number of universes outside of our own each with their own set of physics, realities, and potential life beyond our understanding) and multiple dimensions beyond space and time that our senses and technologies are unable to detect. To understand that there is much more that we don't know and perhaps ever will know than we do know is to begin to adopt a position of humility in contrast to the earlier assertions that Earth was the center of everything and the single focal point for God's creation. The new paradigm is a position of humility and a discarding of postulations and theories premised on our limited perspective from Earth.

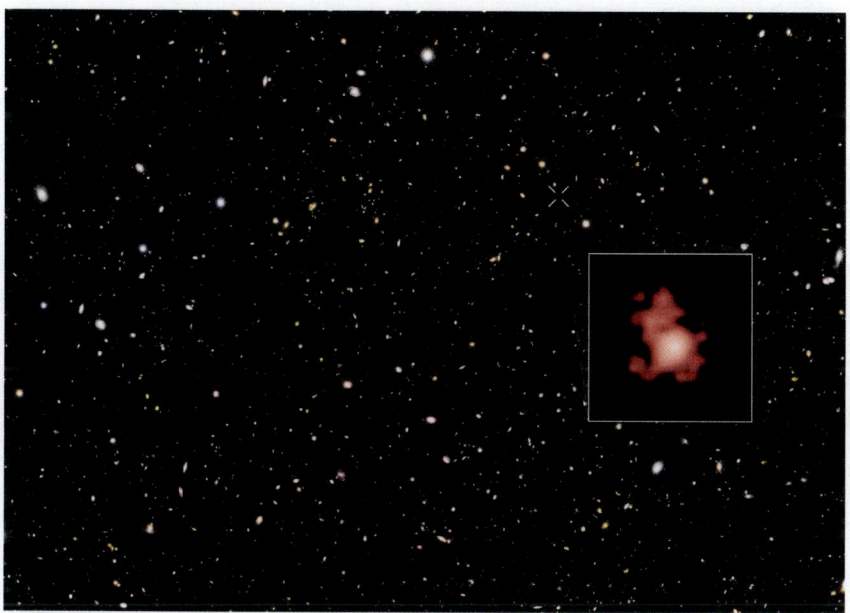

THE FURTHEST KNOWN GALAXY GN-Z11 at 13.4 billion light-years distant as imaged by the Hubble Telescope

The Limits of Time – Where to Draw the Line?

The same progression in our understanding of space is also true for our understanding of time. Some theologians have historically postulated that the world is no more than about 25,000 years old based on interpretation of Scripture. Scientific evidence (i.e., fossil evidence), sets the time frame for humans on Earth at as much as 2,000,000 years. By contrast, the age of the Earth is estimated to be 4.5 billion years and the age of our Universe is estimated to be 13.8 billion years. If our Universe of galaxies continues to expand and accelerate beyond the point of return (the time when the expansion reverses and collapses due to gravity), then eventually stars making up all the galaxies will burn out and the Universe will move towards a vast empty cold dark place. Earth would, of course, have been long gone by this time as limited by the lifespan of our Sun (estimated to be another approximately 4.5 billion years from now). But is time therefore defined as starting with the Big Bang and ending with the point at which space effectively becomes a cold dark place?

To answer this question, first consider that our collective information may not be correct based on our first premise that there is much more that we don't know than we do know. There is still the possibility that our speculations about dark energy could be proven wrong, and the Universe could someday collapse and the Big Bang repeated indefinitely. If this were the case then time, as defined by the collective series of Big Bangs, could be infinite.

A second possibility, even if our Universe does continue expanding, is that there are in fact an infinite number of other universes (multiverses), or other structures, objects, life forms beyond our understanding and imagination. In this second case, time could also be infinite as defined by the collective activity of the entire Cosmos, where the Cosmos is defined as all that is or ever was, and inclusive of potential multiverses.

A third possibility is that there are other dimensions we are not aware of in which the life and death of parallel existences or multiverses could play out. The possibilities are limitless and time could again be infinite.

The net result for us is to reinforce our position of humility in consideration of our unimaginably short history on Earth as compared with the potentially infinite history of the Cosmos.

Random Chance or Intelligent Design?

Considering all that we don't know with regard to space and time what does this say about why we are here and what, if any, meaning we ascribe to our existence in the Cosmos? First and foremost, it reinforces our position and paradigm of humility with respect to our place in the Cosmos in contrast to discarded postulations putting the Earth at the center of everything. In a very profound way, we are not unlike microbes that go about their lives without a conscious understanding of the larger reality that surrounds them. As compared with the diameter of the observable Universe (estimated to be 93 billion light-years), and with an everlasting Cosmos, we are in a very real sense momentary masters of a mere speck of dust for no more than an instant in time in a much vaster reality. Humility is perhaps our greatest reality and the true beginning of understanding.

As we point our most advanced telescopes and instruments into deep space, we continue to unravel the secrets of our Universe whose indescribable beauty is perhaps best left for description by poets. In terms of answering the question of why we are here, there are ultimately three possibilities that are discussed including (1) we exist because of a random set of serendipitous events, governed by the laws of nature, that resulted in the Universe we observe and in life on Earth, (2) we are here as a result of intelligent design suggesting an intelligent Designer or Creator who like the Cosmos ultimately surpasses our understanding, or (3) both are true.

As with the ancient Greeks who believed in the power of reason, many today ascribe to the first possibility whereby we trust in our ability to reason, understand, and explain the evolution of the Cosmos ending with life on Earth as we experience it today. This is the realm of the scientific method whereby our understanding is shaped through direct observation. Through logic and observation, the evolution of the Cosmos is pieced together until the hypothesis lines up with the collective set of observations. Life on Earth is presented as the rise of species with competitive advantage that go on to evolve, whereas those that are unable to compete go extinct. This explanation does not require intelligent design, unless through direct observation the existence of such a Designer or Creator is confirmed.

For the second possibility, that of intelligent design, many also believe that the collective observations of the Cosmos that continue to unveil the seemingly infinite dimensions of space and time, and even the word Cosmos itself which implies order to the Universe that would otherwise move towards chaos, are themselves evidence for such a Creator beyond our understanding. Moreover, the complex nature of the sequence of events that results in our existence, such as the development of an unborn child, point to intelligence and not random chance. Of the complexity and miraculous nature of the human eye, Leonardo de Vinci once wrote:

"What language can express this marvel? Certainly none. This is where human discourse turns directly to the contemplation of the divine."
— Leonardo de Vinci

In the search of the Cosmos for evidence of intelligent design using our best telescopes and instrumentation, perhaps all that is discovered is that the Cosmos is vast beyond comprehension, seems to follow a set of natural laws, and is beautiful with the description best left for poets, and not scientists. Based on the premise of intelligent design, the information gleaned through observation would support a Creator who was similarly described as vast beyond comprehension, is ultimately in control of such natural laws, and who is also best left for description by poets. While observations of the Cosmos could in this sense point to intelligent design, they would not provide insight into any particular purpose we have for being here.

However, there is somewhere else we can look where we may discover our purpose and that may itself point to intelligent design behind our creation. That place is within ourselves, within our own conscious experience in our interaction with our surroundings. Consider the possibility of the moral law – the law of right and wrong written on our hearts and that governs our decisions and actions. Take a simple example of driving on the highway. If someone cuts you off, and we assume in this instance that it is in fact intentional, there are two possibilities for your response. The first possibility is that there is no such thing as a moral law or code and you are troubled by the actions of the other driver because they are interfering with your goals, such as to get to your destination safely or on time. The other possibility is that there is a greater moral law that does not fundamentally vary from person to person, not counting the insane or delusional. With this possibility, consider a third driver on the road who witnesses the behavior of this driver. Now the person cutting you off does not block this third person's goals, as they are safely at a distance, so if there were no moral law they might be indifferent to the incident. However, more than likely, the third person witnessing the incident would also be troubled with the driver's behavior (note that we are not judging the driver, but considering their actions in the moment). Such a reaction from a third person, not affected by the incident, points to a greater moral law, or code of right or wrong behavior, that governs our decisions and actions.

As to why the moral law would be written on our hearts, there are again two possibilities:

- Herd Instinct as biologists would assess the behavior of a herd of elk or a flock of birds who stick together and nurture their young as a direct outcome of their instinctual drive for survival, almost a programmed response to the Cosmos founded in evolutionary theory, and

- Intelligent Design as reflected in the moral law, or code of right or wrong written on our hearts. For this possibility, the moral law would reflect the nature and character of the power or being behind the design. Scratching beneath the surface of such a moral law, a purpose for our lives is revealed. Ultimately, the code of right or wrong is founded in loving others. To love others, we need to treat them as we would like to be treated. If we are driving on the freeway, we treat other drivers with consideration as we would have them do for us. We put the needs of those around us on equal level with our own. Simply stated, intelligent design suggests that the moral law is placed on our hearts as a reminder to love others as ourselves.

In considering whether we are here by chance, or through intelligent design, we might consider whether the love we feel for our spouse, parents, siblings, children, friends, and others is at its core an instinctual, almost programmed, response to our surroundings to enhance our survivability, or whether this love originates from a sincere desire to hope for the best for those around us. Is this love we experience the same as a mother bird caring for her chicks, or as with our ability to reason, is the love such as a mother for her newborn child, special and founded in humility and a deeper purpose to love others as ourselves.

With regard to intelligent design, consider the humble perspective of the psalmist in his words to his Creator:

Psalm 8:4-8
"New King James Version (NKJV)
⁴ What is man that You are mindful of him,
And the son of man that You visit him?
⁵ For You have made him a little lower than the angels,
And You have crowned him with glory and honor.

[6]You have made him to have dominion over the works of Your
hands;
You have put all *things* under his feet,
[7]All sheep and oxen—
Even the beasts of the field,
[8]The birds of the air,
And the fish of the sea
That pass through the paths of the seas."

MOTHER LOON WITH CHICK

MOTHER'S LOVE FOR CHILD

There is also the third possibility, not yet discussed, that both life resulting from the laws of nature, and intelligent design are not necessarily mutually exclusive. Along these lines, consider the words and perspective of astronomer Dr. Robert Jastrow:

> "For the scientist who has lived by his faith in the power of reason, the story ends like a bad dream. He has scaled the mountains of ignorance, he is about to conquer the highest peak; as he pulls himself over the final rock, he is greeted by a band of theologians who have been sitting there for centuries."
> — Robert Jastrow, God and the Astronomers

Each of us decides for ourselves whether we are here with no grand purpose but to get what we can out of life, often at a cost to those around us, or whether there is a greater purpose for our existence – to love one another as ourselves.

CHAPTER 4

ISLAND UNIVERSES

"A person starts to live when he can live outside himself.
Only a life lived for others is a life worthwhile"
— Albert Einstein

"Love one another. As I have loved you, so you must love one
another."
— Jesus Christ

Now close your eyes again and imagine yourself in space. This time,
you see all around you moons, planets, stars, galaxies and all that make
up the Universe we live in. Next, imagine that you are in fact the center
of everything and all that is revolves around you. While this might seem
silly, our nature often times seems to find us interacting with the world
as if we were our own island universe where everything exists to serve
our desires, needs, and goals.

Loving Others as Ourselves

So, if our purpose is to love others as ourselves, we ask how we are
doing in this regard? As each of us lives out our lives there are
opportunities all around us to love others by treating them as we
ourselves would want to be treated. Such opportunities may arise as
others around us are in need, unable to help themselves, but we are in
a position to help, or they may arise as we seek to meet our own needs
and find ourselves in conflict with the needs of those around us.

For the first, opportunities arising from the needs of those around
us, we may come across a family member, friend, or another person
who has a genuine need, but who is unable to meet this need of their
own volition. These are times when someone could use a "helping
hand" and we are in a position to help. This could include something
as simple as offering a smile or friendly hello to someone who seems
down, perhaps changing their outlook on an otherwise gloomy day. It

could be as simple as holding the door for a stranger or choosing to be polite to your waiter or waitress. Often there are those around us who are suffering and through our own sacrifice of time, energy or resources we are able to help, perhaps spending time with someone who finds themselves in a hospital. Opportunities may also present themselves to meet the needs of others through charitable contributions. There are countless examples, but the commonality is looking beyond our own needs to the needs of those around us and reaching out to help.

However, we do not always succeed in choosing to look beyond our own needs. One of the most extreme examples was an incident several years ago when someone was attacked by a thief on a curb resulting in the victim falling unconscious into the street adjacent to the curb. As this incident was caught on video and replayed on news reports, it was apparent that none of the people walking by stepped in to help the individual who remained lying in the street. After a while, a car came around the corner and inadvertently drove over the victim, killing him. Unfortunately, in this instance, no one stepped in to lend a hand, perhaps out of fear or maybe indifference.

In the other instances where we have opportunities to choose to love others, we find ourselves engaged in meeting our own needs, which come into conflict with the needs of those around us. This could be simply traveling on the freeway to a meeting and we make the choice to drive courteously, or to drive as if the road belonged only to ourselves. It could be times when we find ourselves racing to be first in line or to be served, or choosing not to rush to be in front of others. This might also include a time when you come across a purse left on a park bench and decide to seek out its owner. In each instance, our own needs come into conflict with the needs of those around us and we choose to think only of ourselves, or to treat others as we ourselves would want to be treated.

In each of these examples, and in countless other similar circumstances, we make a choice to consider those around us, or we do not. This is not about judging those around us as we often misinterpret what we see or hear, but this is about looking inside ourselves and asking if we would want to be treated in the way that we are treating those around us.

Ultimately as we live out our lives, we each have choices every day to love others as ourselves, or to simply consider our own desires, needs, and goals. The tendency to only consider self is to act as an island universe for which we view everything around us as there to serve us.

Realms of Love

Next consider how we are doing individually, and as a species inhabiting the Earth. Consider four possibilities characterizing the extent to which we individually or as a society live beyond ourselves or for ourselves. These realms of love (the word love applies loosely to the first three) are as follows:

- Realm of isolation,
- Realm of island universe,
- Realm of transactional love, and
- Realm of sacrificial love.

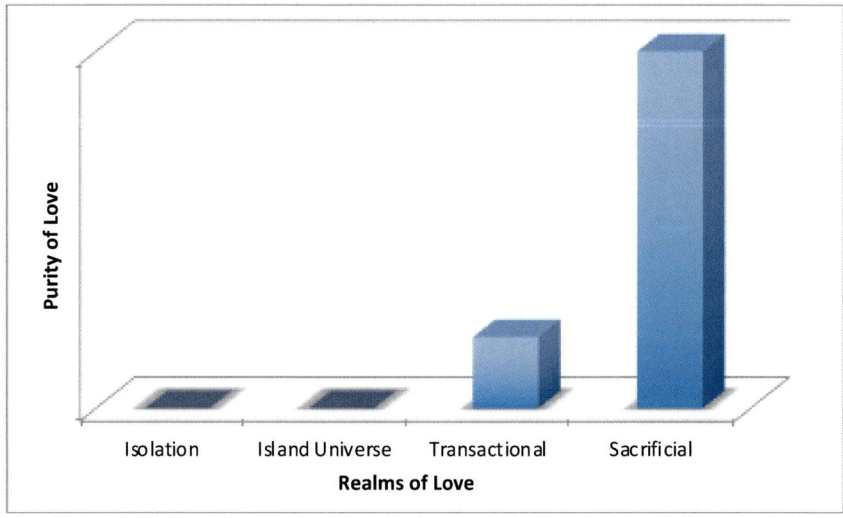

Isolationist Love. In the realm of isolationist love, we may not at the time find ourselves around others with any particular needs, and so we continue on our own path of living out our lives. This is isolation with regard to loving others. An example of this might be when we are spending time at home or engaging in any activity that does not bring us across the paths of other people. Love is neither extended nor are we looking for anything in return from others. The instance in which we are isolated and as a result not in a position to consider others is symbolized in the figure below.

REALM OF ISOLATION

Island Universe. In the realm of island universe, we are experiencing people around us with varying degrees of need, but may decide not to lend a hand, or worse may act in ways that victimize those around us. This is the realm of self-interest and ultimately is the notion of island universes where we are focused on our own desires, needs, and goals, but ignore the same in those around us. There are varying degrees and examples of this realm of love such as:

- criminal behavior where one steals from another or somehow manipulates the circumstances of the situation to their favor, victimizing those around them to meet their own desire, need or goal,

- bullies, prejudice, and chauvinism where one seeks to build themselves up by diminishing those around them,

- indifference where we recognize the needs of those around us but choose to ignore them in favor of our self-interest, and

- ignorance where people around us are in need, but we fail to recognize it and continue down our own path.

The figure on the next page symbolizes the realm of the island universe where we are more focused on ourselves than those around us. In this scenario, we act as if the world and everything in it exists to serve our desires, needs, and goals.

REALM OF ISLAND UNIVERSE

Transactional Love. In the realm of transactional love, we recognize the needs of those around us, and act on that awareness by giving of our time, energy and resources. Again, there are countless examples. Examples might include:

- Doing a favor for someone and expecting something in return,
- Helping out a stranger, and telling everyone what you did, and
- Sacrificing your time, energy and resources for others, but keeping list of your actions for future leverage.

With transactional love we come to the aid of those around us because we expect something in return. In the case of doing a favor for someone, this is purely transactional if we do so only to expect something in return for our efforts.

Another possibility is that we recognize the need of someone around us and step in to address that need which offers nothing material in return, but we find ourselves telling everyone we know what we did. In this instance, it is possible that our actions are motivated by wanting those around us to admire or revere us for our actions, and so the love is not without self-gain.

A third possibility is that we sacrifice our time, energy and resources for those around us and feel a great sense of pride for our actions. While we may not share what we have done with others, we may find ourselves more motivated by wanting to feel this pride than being concerned with the welfare of those around us. We don't tell anyone. However, inside we are still patting ourselves on the back. So even in this example, there is still self-gain at play and the actions are not purely selfless. The figure below symbolizes this third form of transactional love.

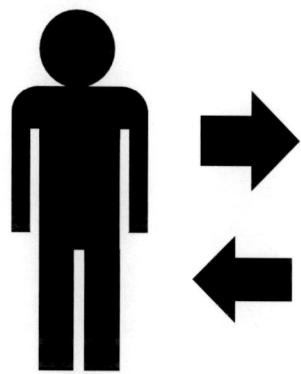

REALM OF TRANSACTIONAL LOVE

Sacrificial Love. Love that at its core is the most selfless usually is at the cost to our own desires, needs, and goals. This is the realm of sacrificial love which offers us all hope for a world where self-interest might otherwise prevail. This hope may be modeled by our parents who may from the time we are born, selflessly pour their time, energy and resources into our general welfare. In time, children may recognize the selfless way in which their parents love them, how that felt as the recipient, and be filled with gratitude. There is perhaps no better way to understand love than to first receive it from others who model this behavior. In receiving love and experiencing gratitude, we are much more likely to ourselves extend authentic unconditional love to those around us. Sacrificial love has the power to influence the course of the lives of those around us. Each one of us can be the agent of change in the world to model sacrificial love and lead others on a similar path. This kind of love aligns with our greater purpose of loving others as ourselves and can bring meaning and purpose to our lives.

In the context of herd theory, as discussed in the previous chapter, sacrificial love can still occur. As a mother wolf cares for her cubs, sacrificing her own time and energy, or as a polar bear does the same for her cubs, these are also examples in nature of sacrificial love.

In the context of the discussion of intelligent design, if we were designed and created to love others as ourselves, and if this reflects the Designer or Creator that is responsible for the gift of life that we have all received, then it would stand to reason that this same Designer or

Creator would also desire to be loved, and would more likely extend the sacrificial love that is at the core of who we are.

Both herd theory and intelligent design in this context are not necessarily mutually exclusive, and both can be true. The figure below symbolizes this fourth realm of love that is focused on the needs of those around us.

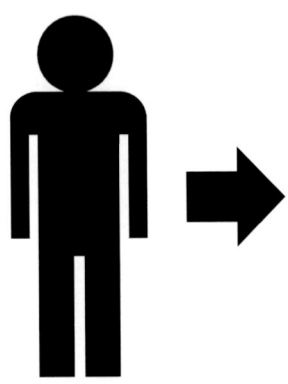

REALM OF SACRIFICIAL LOVE

Love Between Nations - A Candle in the Sky

CRAB NEBULA, M1 in constellation Taurus

During July 1054 in the arid region of what is now northwestern New Mexico, Chacoan stargazers were visited by what may have seemed to some as a candle in the sky. This candle was apparent in broad daylight and outshone Venus in the night sky. This celestial visitor remained visible in daylight for 23 days, but continued to reveal itself in the night sky for almost two years. It was as if the gods had lit a candle, which slowly waned and eventually was extinguished. The Chacoans recorded the event in the form of a petrograph on the underside of a rock cliff shelf below West Mesa in Chaco Canyon. These chance observers may have greeted this new visitor with trepidation, fearing a warning from the gods, and they would not have been far from the mark. Absent the knowledge to unravel the mystery of their guest in the night sky, they may not have grasped that they, along with Chinese, Arabic and Japanese observers elsewhere in the world, were the first to document such a harbinger of the fate of our Sun and fragile home planet.

The myriad of stars in the night sky live out their lives in a manner much resembling our human experience. They are born, spend time as infants, move into adolescence and adulthood, and ultimately enter a time of old age. An old age star that runs out of the fuel it requires to sustain itself faces a demise which varies with the size of the star. Moderate sized stars such as our Sun will first collapse, before then expanding into red giants far beyond their dimension of adulthood. In their expansion, they will consume or scorch any planet in their path, and all that may reside on them, before finally collapsing into white dwarf stars radiating heat, but absent the fuel engine that sustained them through adulthood. A cloud or ring of gas may remain around such white dwarf stars commemorating the star's death and memorializing the fate of any planets that themselves fell victim to their host star's final throes of life.

Stars more than about eight times the mass of our Sun face an end characterized by extreme violence, but that present hope for the Cosmos. As these old age stars run out of fuel, they begin as expansive red giants, not unlike their smaller siblings. However, inside their core a series of contractions occur where the juggernaut of gravity progressively gains the advantage until the core temperature rises upwards of 100 billion degrees Kelvin, and the core recoils and releases a shockwave that in a very short period of time transports the outlying envelop of the star deep into the surrounding space while leaving a remnant neutron star at the center. While violent, such supernovae also serve to deliver the basic ingredients for life (elements forged in the core of the star) into interstellar space and also contribute to the collapse of nearby gas nebulae with the subsequent formation of new solar systems and the potential for new life to arise. Such a supernovae explosion in the northern hemisphere winter constellation of Taurus presented a candle in the sky for the Chacoans in the year 1054.

The message from the gods to the Chacoans is the same for contemporary humans. Our fragile blue planet, this grain of sand that we inhabit in the greater cosmic ocean, is fated to end as our Sun reaches old age. For our species to survive, we must first survive the threat from within that is ourselves, our tendency as individuals and at the national level to act as island universes succumbing to our inward nature that considers self before others. While logic might point to increased consideration of others as our population grows given the increasing opportunity to interact with those around us, we may in truth, if unchecked by our resolve, be challenged in our cities and between nations by increasing division and the potential for alienation. It seems that in the days of small towns, we were more likely to know each other's stories, and our tendency to move outside ourselves began with understanding the stories of those around us, or the nations around us, and we were more likely to move into others' lives with understanding, kindness and compassion. Alienation at the national level, and the advent of nuclear weapons, presents the greatest single risk our species has faced in our time as stewards of this comparatively tiny blue planet we call home.

We must be optimistic and proactive as individuals and as nations to move into the future with hope and promise. As we continue to learn to work collaboratively with other nations we will discover that there is almost no limit to what we can accomplish as a species. In the short term we will face many challenges such as the need for clean energy to fuel our advancing technology, while maintaining the habitability of our home planet. Perhaps we will learn to harness the seemingly limitless energy from the Sun.

In the long term, we as a species will need to move out into interstellar space. While Mars will be good practice for overcoming the challenges associated with habitating other planets, we will ultimately need to travel much further in light of the foreboding fate of the Sun. We will need to voyage into the ultimate wilderness that is the stars themselves. However, doing so depends on our casting away this tendency to act as island universes and love others and other nations as we would ourselves seek to be treated individually and as a nation.

RING NEBULA, MESSIER 57, in constellation
Lyra. Harbinger for the fate of our Sun.

A Reminder from Nature

Imagine that you live in a town where you live out your life day by day, toiling to make a living, rushing to your next appointment, running errands here and there, and seemingly always short on time in a given day to accomplish all that you would like. Perhaps you frequently feel stressed, or short with others around you as a manifestation of your struggle to keep up with life.

Now imagine one day turning a corner on a street that you don't typically frequent and coming across the woods that abut the edge of town. Something about what you see makes you stop in your tracks and momentarily forget why you were in a rush. There is something different in what you see that you had not been aware of a moment ago, and it has you captivated. The woods are thick and the trees are tall. A gentle breeze finds its way through the woods accompanied by the symphony of rustling leaves. At that moment you are connected to nature and reminded that there is more to life than the toil and worry that so frequently consumes your time and energy. Moreover, the visage reminds you of the greater natural world beyond the woods and your perspective changes course as you recall our humble position and purpose in the greater Cosmos. What seemed important moments before now takes a back seat as we are reminded that what is most important are the people in our lives that share this planet, and not the personal goals or material gain that so often is the focus.

SERENDIPITOUS ENCOUNTER WITH THE WOODS

CHAPTER 5

WASTED SPACE

"I believe alien life is quite common in the universe, although intelligent life is less so. Some say it has yet to appear on planet Earth."

— Stephen Hawking

Are We Alone in Our Purpose?

Whether we are alone and the Cosmos is filled with wasted space, or whether it is teaming with life could be considered the subject of science fiction. But many early theories perceived as science fiction were later proven to be factual. Einstein's theory of general relativity was considered by some to be science fiction until later confirmed through experimentation. It is reasonable to consider the conditions to which we attribute our existence, and ask whether these same conditions could occur elsewhere in the Cosmos.

First consider the Earth, and in particular, those conditions or resources that exist and that sustain life. While there are likely many factors to consider, perhaps there is none more important than having liquid water for the origination and sustenance of life. That we have liquid water on Earth may well have provided an optimal medium and shelter for the early development of life. The notion that humans may have originated from life in saline water environments is perhaps evidenced by the fact that our bodies are on the order of 60 percent water, and the salinity of our blood resembles those environments more than freshwater environments. For example, the salinity of human blood is about 0.9 percent (0.9 g salt per 100 g water). For comparison, the following are the average salinities of water environments:

- Seawater at 3.5 percent salinity,

- Estuaries, swamps and brackish seas ranging from 0.05 to 3 percent salinity (brackish water), and
- Freshwater ponds, lakes, rivers ranging from 0 to 0.05 percent salinity.

A second, but not unrelated, factor for the existence and sustenance of life on Earth is the position of the Earth relative to the Sun such that liquid water may be present. The Earth, which is approximately 93 million miles from the Sun, lies within what is referred to as the continuously habitable zone (CHZ), also referred to as the Goldilocks Zone. The CHZ represents the distance from a particular star where a planet would receive adequate light and heat from the star to prevent water from either permanently freezing or evaporating. Generally, the

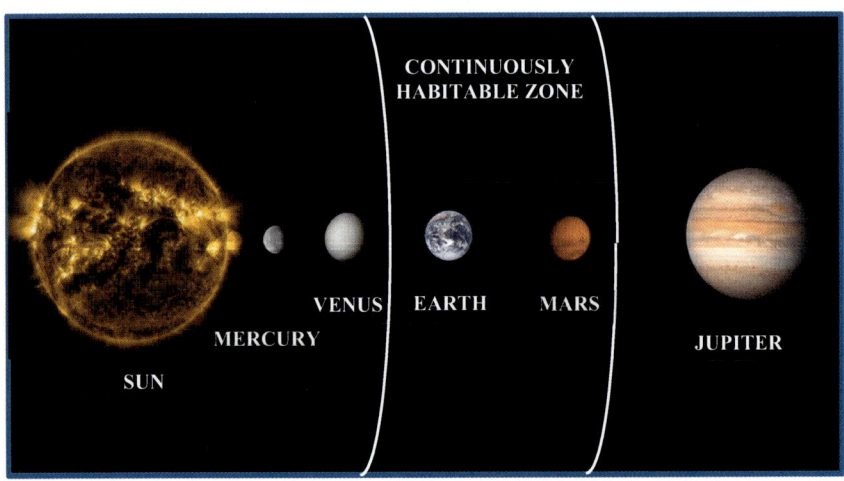

CHZ ENCOMPASSING ORBIT OF EARTH AND MARS

smaller and cooler the star, the closer the planet would need to be for liquid water to be present. Conversely, the larger and hotter the star, the further the planet would need to be for liquid water not to boil off.

Now consider our Sun. To discuss our Sun, it is first necessary to understand how our Sun, a star close-up, compares with other stars. Look at the image of the night sky on the next page and you will notice that the stars are not all white points of light, but rather they vary in color and very generally may be red, yellow, white or blue. Stars vary in their characteristics principally based on their mass, which determines the temperature and color of the star, and predetermines their life cycle. Stellar evolution was described in 1910 by Einar Hertzsprung and Henry Norris Russell and was captured in the so-

IMAGE OF NIGHT SKY revealing stars with different colors

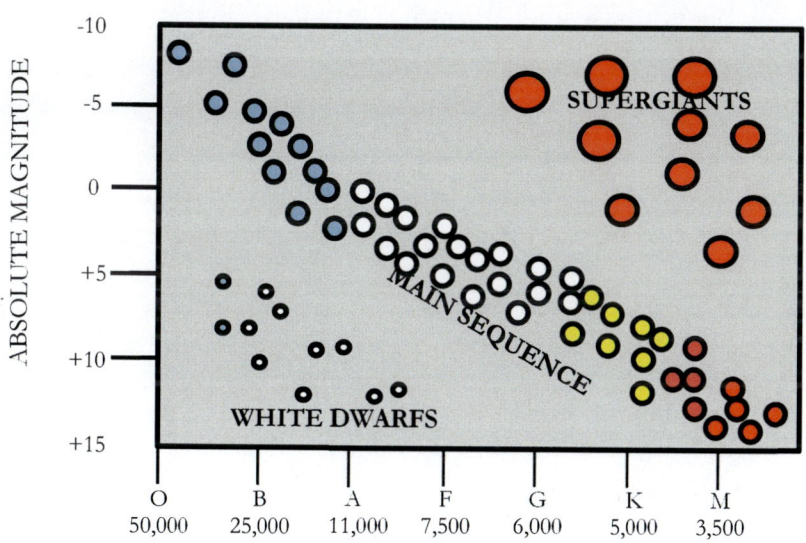

HERTZSPRUNG RUSSEL DIAGRAM

called Hertzsprung/Russel Diagram as shown above. So-called main sequence stars inhabit the central line of stars that span from red dwarf stars on the far lower right, to giant blue stars on the upper left. Stars that are in later stages of life are represented by the other portions of this diagram, inclusive of red giants and white dwarfs. For the purpose

74

of examining star characteristics and those characteristics that might be best in the search for extraterrestrial life, this discussion will focus on main sequence stars.

Red dwarfs, understood to be the most common star in our galaxy, represent stars with a lower overall mass ranging from 0.075 to 0.5 times the mass of our Sun, and consequently burn at a lower surface temperature, around 4,000 K (6,740 F), as compared with other stars. These stars will live for up to 100 billion years because they are using up their fuel slowly; however, because of their small size and low relative heat, they may not be optimal for supporting life-bearing planets. Planets that could sustain life such as on Earth would need to be very close to put them in the Goldilocks Zone, but this might result in the planet becoming tidally locked with respect to the star resulting in great temperature extremes on the planet. Here on Earth we do find examples of extremophiles, bacteria that have found a way to survive under extreme conditions such as those adjacent to deep-sea hydrothermal vents. While there is much we do not know about the Cosmos, our efforts might perhaps be best focused on stars other than red dwarfs.

On the other extreme, blue giant stars are very large ranging from 5 to 10 times the mass of our Sun, and burn at a high surface temperature of 10,000 K (17,540 F). Consequently, these stars burn through their available hydrogen fuel very quickly. Because of the intense heat and short lifespan (on the order of 10 million years) of these stars, they also may not be the best place to focus efforts to search for extraterrestrial life. However, we could be surprised to discover what the possibilities are.

In the central portion of the main sequence lie the yellow and white stars. These stars, of which our Sun is one, tend to be reasonably massive, but not so massive that they burn out too quickly, and not so small as to require planets to be incredibly close in order to receive adequate heat to sustain life.

Other Solar Systems, What Are We Learning?

In recent years, Astronomers have been exploring space for evidence of other worlds outside of our solar system. What they have discovered is no less than phenomenal. Various methods have been used to detect extrasolar planets, and several of those methods are described below.

Direct Imaging. Planets either too close to their star or too far away are either too difficult to distinguish from the nearby star or are too far from the star to reflect sufficient light for imaging in visible wavelengths. With direct imaging, infrared images of a potential star system are taken. Large planets orbiting close to a star tend to have strong infrared signals and so can be imaged in this manner. In addition, the light from the star is sometimes blocked out using an instrument called a coronagraph, which makes detection of nearby planets even more likely. Instrumentation for direct imaging is outfitted on land-based and space-based telescopes for this purpose.

Radial Velocity. With this method, planets are detected indirectly by observing the spectral lines of the host star. If a planet's gravity is affecting the orbit of the star, that is giving the star a small elliptical orbit, then the spectral lines of the star will exhibit a blueshift while moving toward the Earth, and a redshift while moving away from the Earth. For small planets, this method can detect spectral line shifts on the parent star up to 160 light-years away, whereas, for larger Jupiter-sized planets, this method can detect spectral line shifts for stars up to perhaps a few thousand light-years away. Prior to the method of transit photometry, this was frequently the method of choice for those seeking to discover extrasolar planets. A major disadvantage of this method is it limits the study to one star at a time.

Transit Photometry. With this method, a star's brightness is measured, and when a planet crosses (transits) in front of its parent star, the brightness of the star decreases relative to the size of the star and the size of the planet. In contrast to the radial velocity method, transit photometry can evaluate the diameter of the planet. Moreover, while the odds that a particular exoplanet will orbit directly in front of the star relative to our position is low, the method does allow observation of thousands or hundreds of thousands of stars at a time.

With this method in mind, NASA launched the Kepler space observatory in 2009 with its initial mission to scan a large number of stars in the constellation Cygnus (the Northern Cross), and with the hope of finding planets of similar diameter to the Earth. As of June 2018, the Kepler space observatory has identified 2,244 candidate exoplanets, confirmed 2,327 exoplanets, and confirmed 30 exoplanets less than twice the size of the Earth in the Goldilocks Zone around their parent star.

Astrometry. With this method, observations (images) are collected for a particular star over time and the position of the star is accurately recorded. If a planet is present, particularly a large planet with a large orbit, the effect of the planet's gravity will result in the star itself adopting a small elliptical orbit causing its position to change relative to the center of mass over time. A space observatory called Gaia was launched in 2013 by the European Space Agency (ESA), and it is speculated that this observatory will detect thousands of planets during its primary mission of creating the largest ever three dimensional space catalog of approximately one billion astronomical objects.

Exoplanet Discoveries

Exoplanets continue to be discovered in stars within our local neighborhood as shown in the graphic on the next page. Following is a description of some of the recent exoplanet discoveries revealing that our solar system is not unique in regard to the occurrence of planets orbiting their respective host stars.

Alpha Centauri Bb. The nearest star to our solar system, Alpha Centauri lies an estimated 4.36 light-years away, and is found in the skies of the southern hemisphere. As it turns out this is a binary star

ILLUSTRATION OF ALPHA CENTAURI Bb

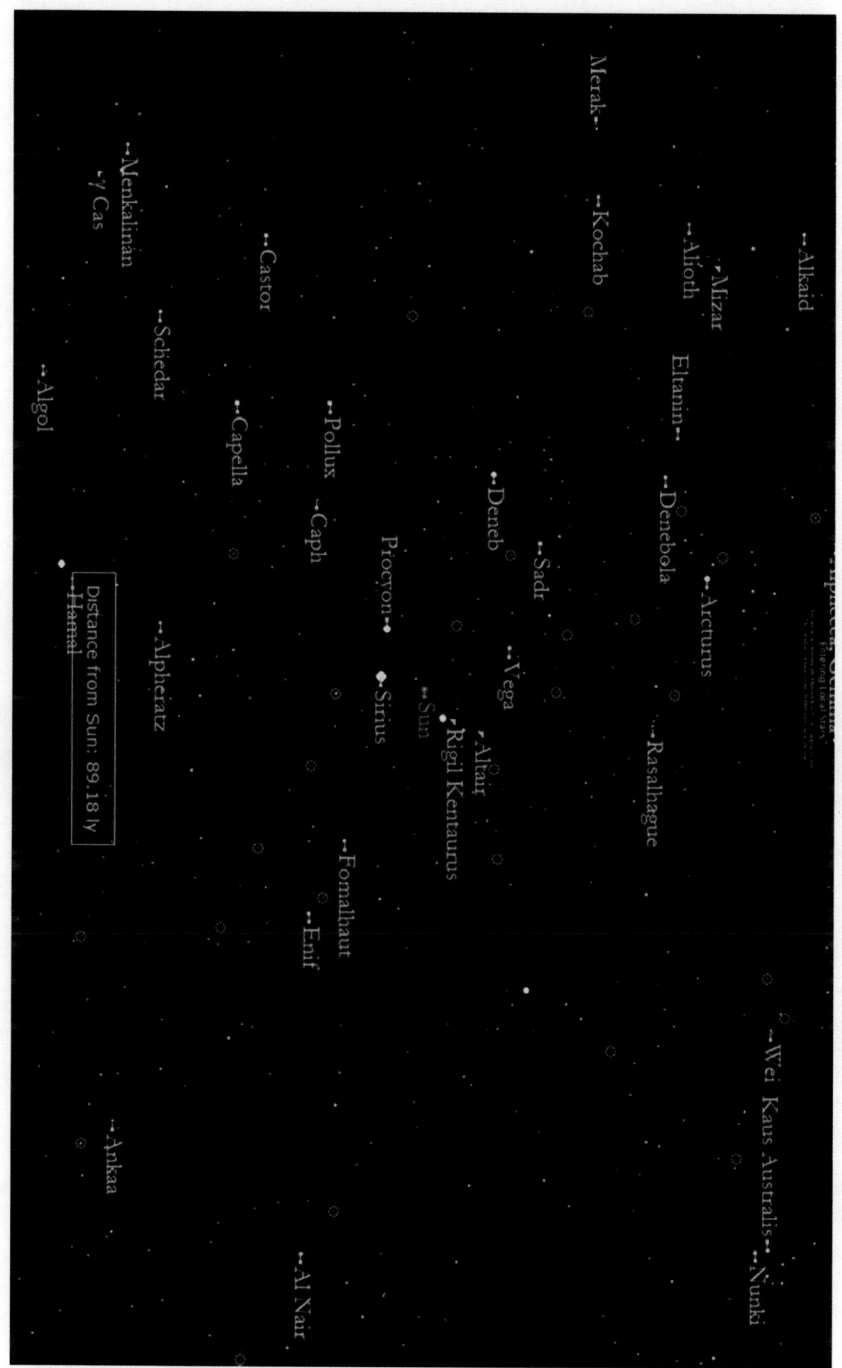

A FEW EXOPLANET DISCOVERIES (red circles) in our stellar neighborhood

system with brighter Alpha Centauri A and dimmer Alpha Centauri B rotating around a shared center of mass. In 2012, Xavier Dumusque of the Geneva Observatory in Switzerland and his colleagues, identified what is reported to be an approximately Earth-sized planet orbiting Alpha Centauri B. This planet is reportedly only about 0.042 AU, or 3.9 million miles, from Alpha Centauri B, and is not understood to be within the Goldilocks Zone. By comparison, Mercury's average distance is 36 million miles from our Sun. If you were to visit this planet, you would likely encounter a desolate surface devoid of life, and with surface temperatures on the order of 2,200 F given its close proximity to Alpha Centauri B. Although the planets dimension appears to be similar to Earth's, it might not have an atmosphere as a direct result of the surface temperatures. Moreover, assuming the planet wasn't tidally locked with Alpha Centauri B based on its close proximity, you would observe a sunrise close-up as Alpha Centauri B would loom large on the horizon, with Alpha Centauri A shining brightly in the distance, and our own Sun just a spec of light fading as the night sky gave way to day under the scorching sunlight.

Kepler 70b. The star Kepler-70, formerly KOI-55, lies approximately 3,849 light-years away in the constellation Cygnus. With an apparent magnitude of only 14.87, this star is not visible with the naked eye. This is an example of a star that has lived through its main sequence life until its hydrogen fuel supply was exhausted. Then around 18.4 million years ago, Kepler-70 expanded considerably as it entered the red giant phase. Subsequently, Kepler-70 contracted to its current state as a subdwarf sustained by helium fusion.

At least two planets were discovered around this star including Kepler-70b and Kepler-70c. Both planets appear to be at a distance from the host star that suggests that they were engulfed by the expanding red giant, and subsequently liberated as the star contracted. Consequently, these planets are likely dramatically different from the time prior to expansion of the star and what remains is likely only the central core of the original planets. For example, Kepler-70b, which lies only 0.006 AU (558,000 miles) from its host star has an estimated mass equivalent to 44 percent of the Earth's mass. The mid-day surface temperature of this planet is thought to be as much as 12,500 F.

ILLUSTRATION OF KEPLER 70b

Fomalhaut b. This 1.16 apparent magnitude star lies in the constellation Piscis Austrinus. Using the Hubble Telescope for direct imaging, astronomers have identified a planet estimated to be 600 times the mass of Earth and that orbits its host star at a distance of about 115 AU (10,695,000,000 miles). Based on the distance from its host star, this planet has an orbital period of 876 years. What makes this star different is that it was detected through direct imaging which tends to discover larger planets further removed from their host star. In contrast, methods such as radial velocity and transit photometry tend to discover fast-moving planets very close to their host stars.

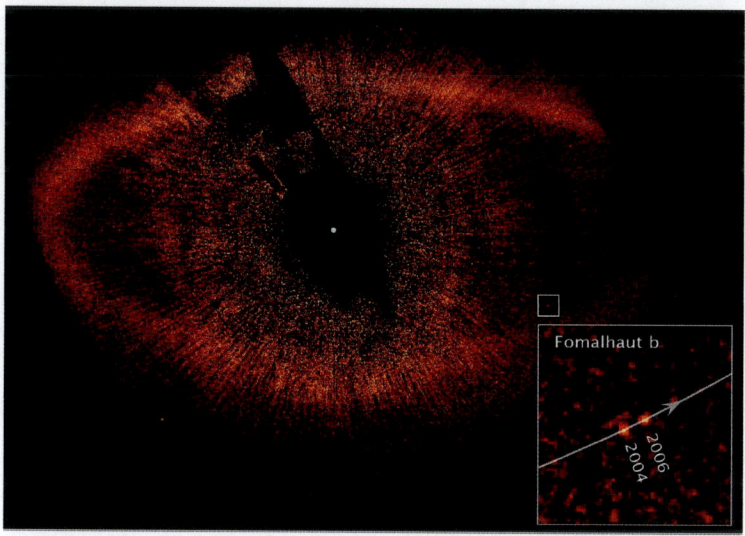

FOMALHAUT B as imaged by the Hubble Space Telescope

PSR B1620-26b. This planet orbits a binary star system in the constellation of Scorpius and consists of a pulsar and a white dwarf found just outside the core of globular cluster M4. The pulsar represents the remnant star from a supernovae explosion, whereas the white dwarf remains from a former sun-like star that exhausted its fuel and ejected its outer layers.

PSR B1620-26b is estimated to be 13 billion years old making it the oldest identified planet as of 2013. It is estimated to be 800 times as massive as the Earth and 2.5 times as massive as Jupiter. It is estimated to lie at a distance of 23 AU (2,139,000,000 miles) and to orbit the binary star system every 100 years.

ILLUSTRATION OF PSR B1620-26b with host stars in distance

WASP-12b. This planet at 1.4 times the mass of Jupiter was discovered at a distance of only 0.02 AU, or 1.86 million miles, from its host star that it orbits every 1.1 Earth days. The host star is in the constellation of Auriga. Hubble Space Telescope was used not only to identify an atmosphere on this planet, but the composition which consists of water, methane, carbon monoxide, and metals such as aluminum, magnesium, tin, and vanadium. However, based on its close proximity to its host star, astronomers believe that this planet's surface temperature approaches 4,700 F. At this temperature, its atmosphere expands to approximately three times its diameter as compared with the thin veneer of our own Earth atmosphere as observed from orbit.

Moreover, the atmosphere of this planet is slowly lost into space and subsequently vacuumed up by its host star such that the planet's atmosphere may only last another 10 million years. Effectively, the host star is slowly cannibalizing WASP-12b.

ILLUSTRATION OF WASP-12b

PSR B1257+12 System. The host star for these planets lies 2,300 light-years away in the constellation of Virgo. The star is known to be a pulsar that is a highly magnetized rotating neutron star that in turn is the collapsed core remnant from a star that has undergone a supernova explosion. Pulsars emit a beam of electromagnetic radiation that is emitted as the star rotates. The host star for these planets rotates every 1.22 seconds. The emission can be detected only when the emission is pointed toward Earth and is very cyclical and predictable.

Based on observations of the pulsar emissions, three planets were discovered to be orbiting this host star as a result of their gravitational tug on the star and the corresponding wobble detected in the star. These planets are located between 0.19 and 0.46 AU from the star with masses that range from 0.2 (about the size of our Moon) to 4.1 times the mass of the Earth. These are understood to be second generation planets which are planets that arose following the death of the star.

They may have originated from the accretion of remnant supernova debris or may have come from a former companion star.

ILLUSTRATION OF PSR B1257+12 SYSTEM

55 Cancri e. This extrasolar planetary system lies in the constellation of Cancer at a distance of 41 light-years away. This host binary star system 55 Cancri consists of a star about the same size as our Sun and a second smaller red dwarf companion star.

Four planets have been identified in this extrasolar solar system, but the most interesting is the one closest to 55 Cancri that is called 55 Cancri e. The planet 55 Cancri e is about 8.4 times the mass of the Earth and lies about 0.016 AU from the host star system. An abundance of carbon has been detected in this star suggesting that both stars and the planets may have formed from an interstellar gas cloud consisting of remnant material from a past supernova explosion. The pressures in such an explosion are thought capable of fusing atoms into heavier elements such as carbon. The closest planet is thought to have a graphite (carbon-based mineral) rich surface underlain by a layer of diamonds several miles thick as diamond can form from graphite (same material as in a No. 2 pencil) subjected to intense heat and pressure such as could occur at depth in a planet.

ILLUSTRATION OF 55 CANCRI e

Kepler-64b. Star system Kepler-64 is a quadruple system with two primary stars (1.5 and 0.4 solar masses) orbiting a common center of mass every 20 days, and two additional stars at a distance of about 1,000 AU (about 30 times the distance from the Sun to Neptune) that seem to be in tow with the first two as they travel through interstellar space.

The planet Kepler-64b (also known as PH1b, which stands for Planet Hunters 1), has a mass of about 169 times that of the Earth (on

ILLUSTRATION OF KEPLER 64b

the order of one half that of Jupiter) and orbits the primary two stars every 138 days at a distance of about 0.65 AU.

Kepler-11 System. Star Kepler-11 is a 14.2 apparent magnitude Sun-like star that lies about 2,150 light-years away in the constellation Cygnus. What makes this star system special is that at least six planets have been identified in orbit around the host star. The planets range in size from 1.9 to 25 times the mass of the Earth. Moreover, all six planets lie within a distance that as compared with our solar system would put them all inside the orbit of Venus.

Extrasolar planetary systems such as Kepler-11 give hope that there may be many such systems with multiple planets such as in our own system and which correspondingly increases the chances of finding planets in the habitable zone.

ILLUSTRATION OF KEPLER-11 SYSTEM

Gliese 667C System. Last but most certainly not the least, this triple star system lies about 23 light-years away in the constellation Scorpius (just above the tip of the tail). The system consists of three relatively small stars, the smallest of which, Gliese 667C, is a red dwarf that is a little less than half the diameter of our Sun and about half the surface temperature of our Sun.

As it turns out, six or maybe seven planets have been found to orbit Gliese 667C, three of which lie within what would be the habitable zone. The habitable zone for such a small star with a relatively low temperature would lie fairly close to the star. As it turns out, three of the planets have been identified at distances ranging from 0.05 to 0.55 AU, which for this star would be within the habitable zone and equivalent to being within the orbit of Mercury for our solar system. Each of the planets is larger than Earth with masses ranging from 2.7 to 6.9 times that of Earth.

This discovery gives much hope to scientists and astronomers hoping to find life elsewhere in the Milky Way Galaxy given that red dwarf stars make up the majority of stars in our galaxy. This also challenges our understanding in that in our initial consideration, we ruled out looking for life in smaller red dwarf stars. However, as with life that finds a way in deep ocean geothermal vents on Earth, the Cosmos is filled with possibilities that we may not have considered.

Since the first discussion of CHZs around host stars in 1953, many star systems have been identified other than Gliese 667c around which planets occupy the CHZ. In November 2013, based on results from planetary research by the Kepler space observatory, astronomers estimated that there could be as many as 40 billion Earth-sized planets orbiting the CHZ of their host stars whether the stars were like our Sun, or whether they were red dwarfs as with Gliese 667c.

ILLUSTRATION OF GLIESE 667C

Exoplanet Implications for Our Galaxy

An equation for intelligent life

In 1961 as preparation for the first scientific meeting for the search for extraterrestrial intelligence (SETI), Dr. Frank Drake wrote what is now referred to as the Drake Equation for the purpose of stimulating a discussion on the subject at the meeting. The Drake equation is shown below.

The Drake Equation

$$N = R* \cdot fp \cdot ne \cdot fl \cdot fi \cdot fc \cdot L$$

N = Number of civilizations in our galaxy whose electromagnetic emissions are detectable

R*= Rate that stars form that are suitable for the development of intelligent life

fp = Fraction of those stars that have planetary systems

ne = Number of planets in each solar system that have an environment suitable for life

fl = Fraction of suitable planets where life actually appears

fi = Fraction of life-bearing planets where intelligent life emerges

fc = Faction of civilizations that develop technology that releases detectable signs of their existence into space.

L = Length of time such civilizations release detectable signals into space.

Different scientists will assign differing values to each of these parameters and will arrive with differing results. As we continue to learn more about the Cosmos, we may narrow down the range of values that apply in this equation. However, as originally estimated by Drake and his colleagues in 1961, the following parameters and results were suggested.

R*= (1 year)$^{-1}$ (one star formation per year on average)

fp = 0.2 to 0.5 (one fifth to a half of all stars formed will have planets)

ne = 1 to 5 (assuming stars identified with planets would have from 1 to 5 planets where life could develop)

fl = 1 (all of these planets will develop life)

fi = 1 (all of these planets will also develop intelligent life)

fc = 0.1 to 0.2 (10 to 20 percent of these will be able to communicate)

L = 1000 to 100,000,000 years

The result, that is the number of civilizations in our galaxy whose electromagnetic emissions are detectable, ranges from a minimum of 20 to a maximum of 50,000,000. As concluded in the SETI meeting, the number was probably on the order of 1000 to 100,000,000 civilizations in the Milky Way Galaxy.

Another way to consider whether or not we are currently alone in our galaxy is to approach the question from the standpoint of our understanding of the circumstances under which intelligent life could develop given the number of stars in our galaxy, regardless of whether or not they develop radio astronomy. One way to consider this question is with the alternate equation below.

$N = ns \cdot fc \cdot fp \cdot fh \cdot fl \cdot fi$

N = Number of civilizations in our galaxy
ns = Number of stars currently estimated to be in our Milky Way Galaxy.
fc = Fraction of those stars that are white or yellow
fp = Fraction of the stars which have planetary systems
fh = Fraction of solar systems with planets in the CHZ
fi = Fraction of planets in the CHZ where life develops
fi = Fraction of life-bearing planets where intelligent life emerges

We will make the following assumptions to estimate the number of civilizations "N".

ns = 250 billion stars based on extrapolation from what can be observed
fc = 0.1 of stars being either white or yellow such that the stars have reasonably long lifespans allowing life to develop and flourish
fp = 0.6 to 0.8 of these stars estimated to have planetary systems
fh = 0.1 to 0.5 of these planetary systems with planets inside the CHZ
fl = 0.001 to 0.1 of these planets in the CHZ where life actually develops
fi = 0.001 to 0.1 of these planets where intelligent life develops

Based on these assumptions, the total estimated number of planets with intelligent life, which we will call a civilization, in our Milky Way could range from 1,500 to 100,000,000.

If we instead are inclusive of red dwarf stars, being the most abundant star in our galaxy (on the order of 85 percent), and where many planets have already been discovered, we then have the following revised assumptions and estimate for number of civilizations "N".

ns = 250 billion stars based on extrapolation from what can be observed fc = 0.95 of stars being white, yellow or red dwarf such that the stars have reasonably long lifespans allowing life to develop and flourish

fp = 0.6 to 0.8 of these stars estimated to have planetary systems

fh = 0.1 to 0.5 of these planetary systems with planets inside the CHZ

fl = 0.001 to 0.1 of these planets in the CHZ where life actually develops

fi = 0.001 to 0.1 of these planets where intelligent life develops

Based on these assumptions, the total estimated number of planets with intelligent life in our Milky Way could range from 14,250 to 950,000,000.

If there are 100 billion galaxies in the visible universe, as estimated based on Hubble Telescope Research, then the total number of such civilizations in the visible universe could be between 1,425,000,000,000,000 and 95,000,000,000,000,000,000, inclusive of red dwarf stars and assuming loosely that other galaxies have similar star composition to our own galaxy. If this is true, the Universe could be teeming with life and our greatest challenge would be establishing communication or traveling over the vast incomprehensible distances of interstellar and intergalactic space. While this seems impossible and unlikely, there is no doubt that it once seemed that only the gods could travel to the Moon and we overcame what seemed impossible. When we consider the possibility of multiverses, the numbers become infinite. Now the notion that the Cosmos is a lot of wasted space is looking less gloomy and far more optimistic.

The Cosmos holds may secrets that our species may never uncover or resolve, in part as a result of our limited vantage point as we have been constrained to observe and listen from our solar system. There are other mysteries in the Cosmos that we may not unravel because we haven't had the imagination to ask the right questions.

In August 2012, mankind took the first step into the impossible as the 1977-launched Voyager I spacecraft traversed the outer boundary of our solar system and entered interstellar space. As of February 2018, Voyager I was about 13.2 billion miles away and still carrying a message from our species to an intelligent civilization that may someday intercept the unmanned spacecraft. By the time this message is intercepted, assuming the spacecraft is not destroyed beforehand, it is uncertain whether our civilization will still exist on Earth, will have inhabited Mars or perhaps one of the moons in our solar system, or whether our species will have begun the cosmic journey into the wilderness that is interstellar space.

IMAGE AND OTHER CREDITS

Note that for all text, quotes, image and other credits that inclusion herein does not in any way constitute an endorsement of this work.

Chapter 1 Home Among the Stars

Page 5. NASA/Apollo 17 crew; taken by either Harrison Schmitt or Ron Evans – Earth https://web.archive.org/web/20160112123725/http://grin.hq.nasa.gov/ABSTRACTS/GPN-2000-001138.html (image link); https://www.nasa.gov/multimedia/imagegallery/image_feature_329.html

Page 7. Mercury by NASA/Johns Hopkins University Applied Physics Laboratory/Carnegie Institution of Washington. Edited Version of Image: Mercury in color - Prockter07.jpg by Papa Lima Whiskey. NASA/JPL, Public Domain, https://commons.wikimedia.org/w/index.php?curid=4163406

Page 8. Venus by NASA/Johns Hopkins University Applied Physics Laboratory/Carnegie Institution of Washington - http://photojournal.jpl.nasa.gov/catalog/PIA10124, Public Domain, https://commons.wikimedia.org/w/index.php?curid=2450357

Page 9. Venus surface. Image Credit: NASA/JPL. *Editor: NASA Content Administrator* https://www.nasa.gov/multimedia/imagegallery/image_feature_358.html

Page 10. Mars. NASA. Editor: NASA Content Administrator. https://www.nasa.gov/multimedia/imagegallery/image_feature_83.html

Page 12. By NASA/JPL – Mars surface, http://photojournal.jpl.nasa.gov/catalog/PIA02405, Public Domain, https://commons.wikimedia.org/w/index.php?curid=2154208

Page 13. By NASA/JPL/USGS - This image (of Jupiter) was catalogued by Jet Propulsion Laboratory of the United States National Aeronautics and Space Administration (NASA) under Photo ID: PIA00343., Public Domain, https://commons.wikimedia.org/w/index.php?curid=11829

Page 14. By NASA, Caltech/JPL – Jupiter Atmosphere http://www.jpl.nasa.gov/releases/2002/release_2002_166.htmlhttp://photojournal.jpl.nasa.gov/catalog/PIA01384 (image link), Public Domain, https://commons.wikimedia.org/w/index.php?curid=12341

Page 15. By NASA/JPL – Saturn, http://photojournal.jpl.nasa.gov/catalog/PIA02225, Public Domain https://commons.wikimedia.org/w/index.php?curid=39233892

Page 18. Titan. By NASA - Vistas del Sistema Solar Procesamiento de imágenes con Paint.NET, Public Domain, https://commons.wikimedia.org/w/index.php?curid=39276844

Page 19. By NASA/JPL-Caltech – Uranus https://www.nasa.gov/feature/jpl/voyager-mission-celebrates-30-years-since-uranus

Page 21. By NASA - JPL image, (Neptune), Public Domain, https://commons.wikimedia.org/w/index.php?curid=640803

Page 22. By NASA / Johns Hopkins University Applied Physics Laboratory / Southwest Research Institute – Pluto http://www.nasa.gov/sites/default/files/thumbnails/image/crop_p_color2_enhanced_release.png (Converted to JPEG) (see also PIA19952}, Public Domain, https://commons.wikimedia.org/w/index.php?curid=41837276

Page 24. Graphic of local stars from Deep Space Explorer software, 2001 available at Space.com. From the creators of Starry Night.

Page 25. Alpha Centauri Image credit: ESA/NASA, Editor: Ashley Morrow, https://www.nasa.gov/image-feature/goddard/2016/hubbles-best-image-of-alpha-centauri-a-and-b

Page 26. Sirius. By NASA, ESA and G. Bacon (STScI) -
http://www.spacetelescope.org/images/heic0516b/, Public Domain,
https://commons.wikimedia.org/w/index.php?curid=477456

Page 27. Image of Milky Way by Edward Tyler

Page 28. Milky Way location of Sun. Editor: Holly Zell. NASA.
https://www.nasa.gov/mission_pages/sunearth/news/gallery/galaxy
-location.html

Page 29. By NASA/JPL-Caltech/UCLA – Andromeda Galaxy
http://www.nasa.gov/mission_pages/WISE/multimedia/pia12832-
c.html, Public Domain,
https://commons.wikimedia.org/w/index.php?curid=9531631

Page 29. Whirlpool Galaxy image by Edward Tyler

Page 30. Graphic of local group of galaxies from Deep Space
Explorer software, 2001 available at Space.com. From the creators of
Starry Night.

Page 32. Hubble Ultra Deep Field. By NASA and the European
Space Agency. –
http://hubblesite.org/newscenter/archive/releases/2004/07/image/
a/warn/, Public Domain,
https://commons.wikimedia.org/w/index.php?curid=1499793

Chapter 2 Time Machines

Page 35. Table for days of week.
https://en.wikipedia.org/wiki/Names_of_the_days_of_the_week

Page 40. Image of Mizar by Edward Tyler

Page 41. Image of Orion Nebula by Edward Tyler

Page 42. Andromeda Galaxy. By NASA/JPL-Caltech - NASA, Public
Domain,
https://commons.wikimedia.org/w/index.php?curid=19525315

Page 43. Image of Messier 61 by Edward Tyler

Page 44. Einstein Photograph by Oren Jack Turner, Princeton, N.J.
(The Library of Congress) [Public domain], via Wikimedia Commons

Page 45. Graphic of special relativity and time dilation created by
Edward Tyler

Chapter 3 Cold Dark Space

Page 48. Image of Messier 81 by Edward Tyler

Page 49. NASA. Wilkinson Microwave Anisotropy Probe (WMAP) team, Baby Universe. Public Domain, https://commons.wikimedia.org/w/index.php?curid=29266

Page 50. Universe image from Deep Space Explorer software, 2001 available at Space.com. From the creators of Starry Night.

Page 51. Galileo. By Justus Sustermans - http://www.nmm.ac.uk/mag/pages/mnuExplore/PaintingDetail.cfm?ID=BHC2700, Public Domain, https://commons.wikimedia.org/w/index.php?curid=230543

Page 51. Image of Jupiter by Edward Tyler

Page 52. By Johan Hagemeyer (1884-1962) – Edwin Hubble http://hdl.huntington.org/cdm/ref/collection/p15150coll2/id/129, Public Domain, https://commons.wikimedia.org/w/index.php?curid=38358266

Page 53. Images of Sombrero Galaxy and Pinwheel Galaxy by Edward Tyler

Page 54. GN-Z11 Hubble Imagery. NASA, ESA, P. Oesch and B. Robertson (University of California, Santa Cruz), and A. Feild (STScI) https://www.nasa.gov/feature/goddard/2016/hubble-team-breaks-cosmic-distance-record

Page 59. Image of mother loon with chick taken by Brant Smith

Page 59. By Henry Essenhigh Corke (1883-1919) - 'Mother and Child', No restrictions, https://commons.wikimedia.org/w/index.php?curid=8497084

Chapter 4 Island Universes

Page 63. Graphic of Realms of Love by Edward Tyler

Page 64. Symbolic image (and related images in chapter) for isolationist love by Edward Tyler

Page 67. Image of Crab Nebula Messier 1 by Edward Tyler

Page 70. Image of Ring Nebula Messier 57 by Edward Tyler

Page 71. Forest image by Edward Tyler

Chapter 5 Wasted Space

Page 73. Graphic of CHZ produced by Edward Tyler using previously credited images, except Sun which is credited to: NASA/SDO, *Editor: Holly Zell*

Page 74. Image of stars by Edward Tyler

Page 74. Hertzsprung Russel Diagram produced by Edward Tyler

Page 77. Alpha Centauri Bb illustration (no changes made). By Jelle Gouw - Own work, CC BY-SA 4.0, https://commons.wikimedia.org/w/index.php?curid=37898174 License link: https://creativecommons.org/licenses/by-sa/4.0/deed.en

Page 78 Graphic of local planets from Deep Space Explorer software, 2001 available at Space.com. From the creators of Starry Night.

Page 80. Kepler 70b. By MarioProtIV - Space Engine v0.9.7.3 (planet located by me), CC BY-SA 4.0, https://commons.wikimedia.org/w/index.php?curid=49101201 License link: https://creativecommons.org/licenses/by-sa/4.0/deed.en

Page 80. By Formalhaut_b.jpg: NASA, ESA, P. Kalas, J. Graham, E. Chiang, E. Kite (University of California, Berkeley), M. Clampin (NASA Goddard Space Flight Center), M. Fitzgerald (Lawrence Livermore National Laboratory), and K. Stapelfeldt and J. Krist (NASA Jet Propulsion Laboratory) derivative work: Jan.Kamenicek (talk) - Formalhaut_b.jpg, Public Domain, https://commons.wikimedia.org/w/index.php?curid=10286117

Page 81. Exoplanet PSR B1620-26b. By Illustration Credit: NASA and G. Bacon (STScI) - http://hubblesite.org/newscenter/newsdesk/archive/releases/2003/19/image/a, Public Domain, https://commons.wikimedia.org/w/index.php?curid=1332124

Page 82. Exoplanet. Wasp-12b By NASA -
http://hubblesite.org/newscenter/archive/releases/2010/15/,
Public Domain,
https://commons.wikimedia.org/w/index.php?curid=10511517

Page 83. Exoplanets PSR B1257+12. By NASA/JPL-Caltech/R.
Hurt (SSC) - http://photojournal.jpl.nasa.gov/catalog/PIA08042,
Public Domain,
https://commons.wikimedia.org/w/index.php?curid=705149

Page 84. Exoplanet 55 Cancri e. By NASA/JPL-Caltech -
https://photojournal.jpl.nasa.gov/jpeg/PIA22069.jpg, Public
Domain,
https://commons.wikimedia.org/w/index.php?curid=64110468

Page 84. Exoplanet Kepler-64b. image Credit: Haven Giguere/Yale
Last Updated: Aug. 7, 2017, Editor: NASA Content Administrator
https://www.nasa.gov/content/kepler-64b-four-star-planet

Page 85. Kepler 11 planetary system. By NASA / Tim Pyle - New
Planetary System image:[1], Public Domain,
https://commons.wikimedia.org/w/index.php?curid=12890676

Page 86. Gliese 667C illustration. By ESO/L. Calçada - ESO, CC BY
4.0, https://commons.wikimedia.org/w/index.php?curid=21693141

REFERENCES

Astronomy Magazine. October 2013 issue.

Ferris, Timothy. *Galaxies.* Steward, Tabori & Chang, Publishers, New York.

Green, Brian. NOVA PBS. *The Fabric of the Cosmos* WGBH Educational Foundation 2011.

Hawking, Stephen. *A Brief History of Time.* Tenth Anniversary Edition. Bantam Books. 1988.

Jastrow, Robert. *God and the Astronomers.* W.W. Norton & Company, Inc. New York/London, 1992.

Lewis, C.S. Mere Christianity. Harper San Francisco, A Division of Harper Collins Publishers, 1952.

Rood, Robert T., and James S. Trefil. *Are We Alone?* Charles Scribner's & Sons, New York. 1981.

Sagan, Carl. *Cosmos.* Random House, New York, 1980.

www.nasa.gov

www.space.com

wikipedia.org

Made in the USA
Monee, IL
06 March 2020